建筑立场系列丛书 No.93

非类型化建筑
Against the Typology

于风军 林英玉 罗茜 张可新 王方冰 | 译
[荷兰] MVRDV建筑设计事务所等 | 编

大连理工大学出版社

建筑立场系列丛书 No. 93

004	奥斯陆中央车站的Fjordporten摩天大楼 _ Reiulf Ramstad Architects + C.F Møller Architects
008	娜茨卡港口 _ Kjellander Sjöberg
012	里程碑 _ MVRDV
016	城市绿脊 _ UNStudio + Cox Architecture

景观中的智利住宅

020	引人入胜的陈列室 _ José Luis Uribe Ortiz
026	Hualle住宅 _ Ampuero Yutronic
042	Loba住宅 _ Pezo Von Ellrichshausen
060	Beach住宅 _ Schmidt Associates Architects
074	Ghat住宅 _ Max Núñez
094	H 住宅 _ Felipe Assadi Arquitectos

学生宿舍
非类型化建筑

112	非类型化建筑 _ Marco Atzori
120	Olympe de Gouges学生宿舍 _ ppa + Scalene + AFA
136	大田大学惠化寄宿学院 _ Seung, H-Sang
156	塔帕尔大学学生宿舍 _ McCullough Mulvin Architects
172	西城国际大学爱家宿舍 _ Studio Sumo
186	卢西安·考尼尔学生宿舍 _ A + Architecture

| 196 | 建筑师索引 |

C3 No. 93 Against the Typology

004 Fjordporten in Oslo Central Station_Reiulf Ramstad Architects + C.F Møller Architects

008 Nacka Port_Kjellander Sjöberg

012 The Milestone_MVRDV

016 Green Spine_UNStudio + Cox Architecture

Chilean Houses in the Landscape

020 Cabinet of Curiosities_José Luis Uribe Ortiz

026 Hualle House_Ampuero Yutronic

042 Loba House_Pezo Von Ellrichshausen

060 Beach House_Schmidt Associates Architects

074 Ghat House_Max Núñez

094 H House_Felipe Assadi Arquitectos

Student Housings
Against the Typology

112 Against the Typology_Marco Atzori

120 Olympe de Gouges Student Housing_ppa + Scalene + AFA

136 Daejeon University Hyehwa Residential College_Seung, H-Sang

156 Thapar University Student Accommodation_McCullough Mulvin Architects

172 Josai International University iHouse Dormitory_Studio Sumo

186 Lucien Cornil Student Residence_A+Architecture

196 Index

多用途城市大楼 Urban Mixed-use Tower

奥斯陆中央车站的 Fjordporten 摩天大楼
Fjordporten in Oslo Central Station _ Reiulf Ramstad Architects + C.F Møller Architects

由Reiulf Ramstad Arkitekterling牵头，并与C.F Møller Architects、Bollinger + Grohmann Ingenieure、Baugrundinstitut Franke-Meißner und Partner, GMBH以及Transsolar Climate Engineering 等合作的设计团队创作的"北欧之光"在奥斯陆中央车站的Fjordporten摩天大楼设计竞赛中胜出。

该设计竞赛由Bane NOR Eiendom (BNE) 房地产公司举办，于2017年6月邀请了四支设计团队参与这次有很多限制性条款的规划和建筑设计竞赛。评委会根据竞赛方案和评估标准，在全面酌情评估的基础上，评审了参与竞赛的设计提案。在所有提案中，"北欧之光"入选为最佳方案。"北欧之光"设计方案通过可持续性的建筑设计热情地寻找解决方案，从而应对项目所带来的挑战，圆满完成了这项设计任务。

考虑到酒店的数量持续增多，以及歌剧院、德克曼斯克图书馆和Barcode大楼相距很近，该项目设计方案与火车站和附近地区的总体规划进行了良好的结合。该设计方案想要将挪威木结构建筑的传统与结构混凝土和高科技玻璃等现代材料相结合。预计这将有助于修复和加固被保护的Østbanen结构，并让它成为车站未来的视觉标志。因此，它将是一个高效而愉快的会议场所，为游客提供全新的优质空间体验。

同时，为了成为一个适合人类生活和对环境友好的"灯塔"，项目设定了雄心勃勃的可持续性发展目标，还结合了英国建筑研究院环境评选 (BREEAM) 的最高标准，充分考虑了寿命周期评估和成本 (LCC、LCA)，实现了建筑的灵活使用，并以此聚焦项目的实施与其未来的长远发展。

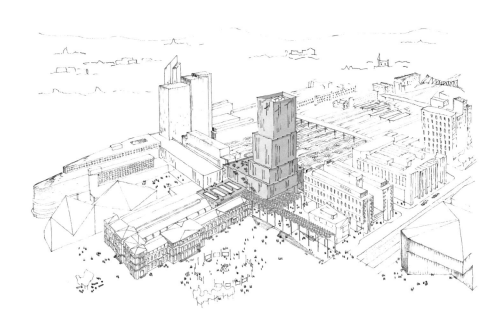

Bane NOR集团将在未来几年重新开发规划奥斯陆市，重新开发规划奥斯陆中央车站和Fjordporten大楼只是第一步。随着新的"北欧之光"项目的展开，Fjordporten大楼将以它所处的中央位置和高大的体量在城市景观中发挥重要作用，并成为集交通、旅游、办公、文体娱乐等功能为一体的枢纽场所。

"Nordic Light" has been awarded as the winner for the design competition of "Fjordporten Oslo S".
The winning team consists of Reiulf Ramstad Architects in collaboration with C.F Møller Architects, Bollinger + Grohmann Ingenieure, Baugrundinstitut Franke-Meißner und Partner, GmbH and Transsolar Climate Engineering.

The competition was held by Bane NOR Eiendom (BNE) who invited four teams in June 2017 to design a restricted planning and architecture. The jury has considered the proposal on the basis of a comprehensive discretionary assessment, contingent on the competition program and the evaluated criteria. Among the entries, "Nordic Light" was chosen to be the proposal that best responds to the program's challenges and the task's premises while enthusiastically finding solutions through sustainable architecture.
The project showcases a good integration with the station and the overall blueprint of the nearby areas, considering the growing density of hotels, the proximity of the Opera,

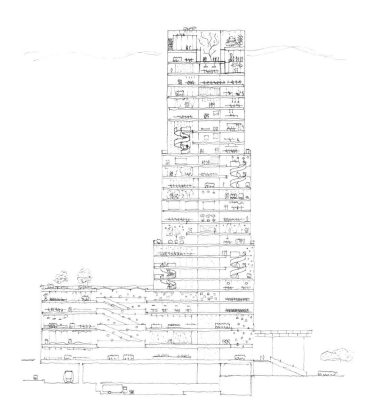

the Deichmanske Library and the Barcode. It wishes to bring back the tradition of Norwegian timber architecture in combination with modern materials such as structural concrete and high-tech glass. This is expected to help the rehabilitation and enhancement of the protected Østbanen structure and its visual identity. Thus, it will be a highly efficient yet pleasant meeting arena that offers the travellers great new spatial and qualitative experiences.

At the same time, to become a human- and environment-friendly "lighthouse", it sets ambitious sustainability goals, combining high BREEAM classification (excellent-level), as well LCC, LCA, flexible solutions of the building utilization, and focuses on the implementation of the project and the upcoming operating period.

Bane NOR will develop Oslo S in the coming years and the first step in this enterprise is the development of the station and the "Fjordporten" tower in Oslo S. With the new "Nordic light", Fjordporten will have a prominent role in the cityscape with its central location, its volume, height, and become the collective hub where for transport, travel, work, leisure, and culture are merged with the site context.

娜茨卡港口
Nacka Port _ Kjellander Sjöberg

Kjellander+Sjöberg建筑设计事务所赢得了位于娜茨卡市和斯德哥尔摩市之间的娜茨卡港口的设计竞赛。娜茨卡港口是一个新城区。之前，娜茨卡市政府把它定位为一个独特地标的合适位置。该街区将通过一个引人入胜、富有多样性的项目为城市环境增添活力。

娜茨卡港口也被称为Klinten街区，它处于Sickla Köpkvarter商业中心和Hammarby Sjöstad区之间的多元化城市环境中，不仅拥有群岛环境，还有许多较古老的工业建筑。

该设计竞赛项目要求解决如何创造有价值的当地体验这一难题，并且设计出独特而有意义的建筑，能够突出该地区的地区特性、节点和路径，同时改善城市环境之间的互通互联。该获胜设计方案以Klinten油漆厂周围历史悠久的工业环境作为出发点，打造出新型的城市空间，让具有当地特色的城市生活更加令人心驰神往。

Klinten街区既是居民住宅区，也是旅游景点。它是一个聚集场所，一天24小时提供各种活动，有集市、餐厅、自行车咖啡馆、共享办公空间、酒店和健身房，也有艺术家工作室、工作坊和公共活动区，反映出该城市的丰富多样性。该街区的设计旨在促进当地的城市生活，打造一个让人可以像在家里一样轻松地进行自由自主活动的地方，鼓励居民利用户外环境进行共创，或者仅仅用来会友和社交。

该项目在整个街区打造了一系列灵活而开放的广场，这些广场斜穿街区，具有多样化的特征，可以吸引居民到此进行各种活动，既可以开办集市，也可以开露天咖啡馆。它还面向周围社区提供了一定的空间，让过路行人和骑自行车的人可以在这停留。整个街区由两部分组成，下面较低部分面向相邻的交通道路，而西南方向是一个风格一致的整体体量，由两座高度不同的塔楼组成。这两座塔楼的交错立面面向南面和未来的公园。沿着建筑交错的两侧产生的后撤空间，包括适合种植和绿化的公共和私人露台。整个街区设计从不同角度和方向提供了不同的视野，创造出一种三维的流畅体验。

娜茨卡港口设计理念的核心主要是为满足人类和地球的未来需

求打造一个积极的愿景。这座注重生态服务系统的高层建筑碳排放积极,采用了可再生材料,拥有绿色健康的微气候,并且具有生命周期可适应性。娜茨卡港口将成为一个连接并创造一个鼓舞人心的都市化可持续生活方式的地方。

整个规划过程现已开始,计划在2020年正式落成。

Kjellander Sjöberg has won the competition for Nacka Port, a new urban district, situated in the area between Nacka and Stockholm. The site has previously been identified by Nacka Municipality as a suitable location for a distinct landmark. The block will contribute to a vibrant urban context with an inviting and varied program.

Nacka Port, the block Klinten, is in a diverse urban setting between the commercial centre Sickla köpkvarter and the district of Hammarby Sjöstad, with features of both an archipelago environment and older industrial buildings. The competition program called for a solution to how valuable local experiences can be created, combined with a distinct and meaningful building that strengthens the identity, nodes and paths of the area as well as improving the connections within the urban setting. Using the historically interesting industrial environment around the Klinten paint factory as a point of departure, the winning proposal generates new types of urban spaces, and contributes to a

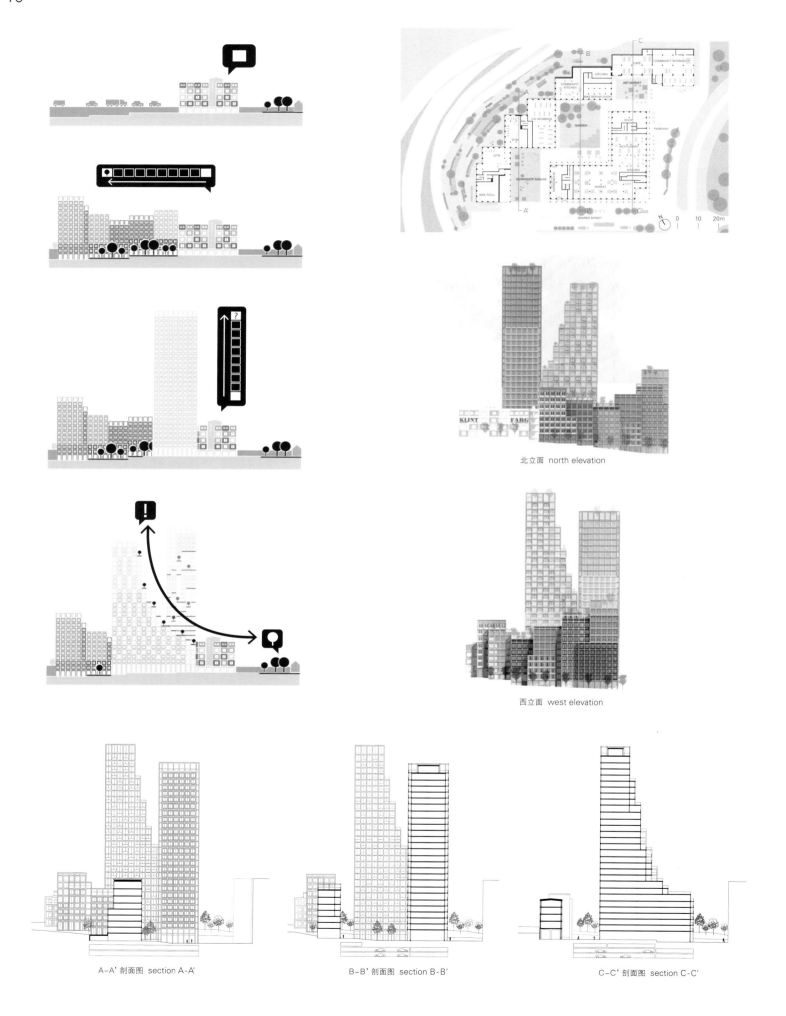

北立面 north elevation

西立面 west elevation

A-A' 剖面图 section A-A'

B-B' 剖面图 section B-B'

C-C' 剖面图 section C-C'

local, inviting city life.

The block Klinten is a destination for residents and visitors. It is a meeting place that offers a wide spectrum of activities during all hours of the day, with a content that mirrors the rich variety of the city; comprising a market, restaurants, a bike-café, coworking spaces, a hotel and gym as well as artist studios, workshops and areas for communal activities. The block is designed to stimulate local urban life, where one feels at home with the freedom to take personal initiatives, where residents are encouraged to use the outdoor environment for co-creation, or just meet and socialize. The project creates a sequence of squares diagonally through the block, flexible and open, with varying characteristics that invite initiatives from residents, markets and open air-cafés. It also provides spaces facing the surrounding neighborhood where pedestrians and cyclists can stop by. The block is composed of a lower part facing the adjacent traffic and a unified volume to the southwest, which divides into two towers of different heights. Two staggered facades face the south and the future park. The setbacks generated along the staggered sides of the building comprise communal and private terraces suitable for cultivation and greenery. The block is designed to provide varied impressions from different directions, a three-dimensional fluid experience.

The core of the Nacka Port concept focuses on creating a positive vision of the future needs of both humans and our planet. A tall building is carbon positive, lifecycle-adapted with renewable materials and a healthy microclimate, focusing on ecosystem services. Nacka Port will be a place to connect and create an inspiring urban and sustainable lifestyle.

The planning process has now begun, aiming to legally bind the development plan in 2020.

多用途城市大楼 Urban Mixed-use Tower

里程碑
The Milestone _MVRDV

德国内卡河畔埃斯林根的新德姆塔特区有着悠久的工业历史,这里曾是货运仓储地。MVRDV建筑设计事务所为这座城市设计了这一崭新的里程碑,而这块"水晶石"将会吸引人们更加关注这座城市及其发展。这座多功能的办公楼曾被称为"Block E",是这片刚刚发展起来的街区中心的标志性建筑,而这片街区未来将会再加入一所大学、住宅开发项目以及众多的零售商店等。MVRDV的目标是建造一座可以体现埃斯林根城市特点的大楼,并且接纳周边的环境,满足使用者的需求。它可以向坐火车经过的人和站在山坡上眺望城市的人发出信号——"我们在这里",它彰显了城市的骄傲、历史以及未来。

这座大楼的立面刻画出埃斯林根城市的边界和自然景观的地形,它是通过一系列构成"埃斯林根空间"的像素图案嵌入环境中的。而一系列的台阶、坡地和平台延伸至大楼的另一侧,以城市中心公共空间的形式形成了一个视角。在这里,游客们可以欣赏葡萄园和附近山丘的景色。

建筑利用内含光伏电池的熔块玻璃组成部分镜面的立面,以映照出周围的环境、城镇、山丘和人。它展现出一幅埃斯林根及其周边地区的像素化的地图。每一个像素都蕴含着不同的信息,讲述着这个城市和它的居民的故事。通过手机App,人们可以探索该城市丰富的内涵,共同创造城市的"公共图书馆"。在地面层,这块水晶石向面前的公共广场打开,将城市与建筑相连接,并且提供了餐厅、咖啡馆和聚会空间等一系列公共设施。建筑的上层是现代化的办公空间,提倡一种健康、平衡的生活和工作关系。

与之形成对比的是,在晚上,发光的立面营造出一种非凡的气氛,使整个建筑成为埃斯林根市的灯塔。建筑前方的广场成为市民们新的聚会场所,也成了城市的下一个里程碑。该项目受德国RVI委托设计,项目计划于2020年开始。

The "Neue Weststadt" in Esslingen am Neckar, Germany is a former freight depot with a diverse industrial history. MVRDV designs a new Milestone for the city, which will draw attention to the town and its developments, a "crystal rock".

Formally known as "Block E", the new mixed-use office building marks the centre of this newly developed district, which will further accommodate a university, housing developments and retail stores. MVRDV's ambition is to generate a building that shows the city of Esslingen and at the same time, opens up to its surrounding and its users. To the people who pass by on the train, and to those that look at the city from the hills, it shouts, "Here We Are", showing its pride, its history, and its future.

The facade traces the boundaries of Esslingen and the topography of its landscape, which is pushed in through a series of pixels that form an "Esslinger Room". A series of stairs, terraces, and platforms emerges and leads to the other side, reaching a viewpoint in the shape of the city centre for public use, where visitors can enjoy the views of the vineyards and surrounding hills.

The facade is designed as partially mirrored, with fritted glass containing PV cells that mirror the environment, the town, the hills and the people. It shows the pixelated map of the area of Esslingen and around. Each pixel carries different information, featuring the stories of the city and its inhabitants. Accompanied by a smartphone app one can discover the richness, creating the public library of the town. On the ground level, the crystal rock opens up to the public square located in front, connecting the city with the building and providing public amenities including a restaurant, café and meeting areas. On the upper levels, modern office spaces are created to encourage a healthy work-life balance.

In contrast, at night, the building becomes a beacon for Esslingen, illuminated through its facade, creating an extraordinary atmosphere. This turns the public square in front into a new meeting point for inhabitants and the city's next milestone. The project was commissioned by RVI Germany. Construction is scheduled to begin in 2020.

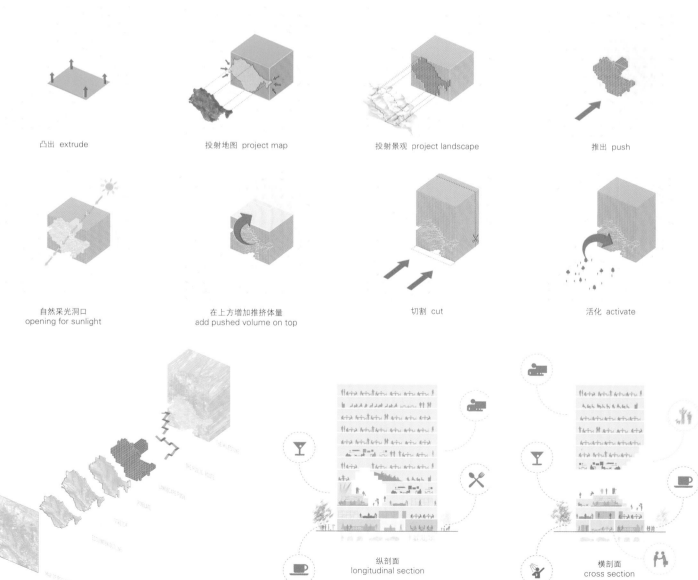

凸出 extrude	投射地图 project map	投射景观 project landscape	推出 push
自然采光洞口 opening for sunlight	在上方增加推挤体量 add pushed volume on top	切割 cut	活化 activate

纵剖面 longitudinal section

横剖面 cross section

城市绿脊
Green Spine _ UNStudio + Cox Architecture

由UNStudio建筑事务所和Cox Architecture建筑事务所共同提交的设计方案"城市绿脊"成为该项目设计竞赛的获奖设计。这座位于墨尔本市中心、预计耗资超过20亿美元的多功能建筑将成为全澳大利亚最高的大楼。

UNStudio的设计方案旨在为南岸地区和墨尔本打造一个新的目的地。整个方案的中心思路是利用一道纵向的"绿脊",将平台、露台和走廊空间连接为一个整体。这一多层面的脊柱是通过一个单体结构从核心向两端分裂而形成的,最终呈现为两个相互分离的超高层大楼,这使得它们在从一个光线充足的峡谷地带上升的过程中,展示了它们核心层的几乎所有地质层。

这一设计手法能够为两侧的大楼带来通透的城市视野,并显著地改善了建筑与周边大环境的连接。大楼中的住宅、办公空间和酒店均能享受到充足的日光和室外空间。

"城市绿脊"的朝向使群楼的公共区域得到了进一步扩展,使绿色空间一路延伸至大楼,有利于建筑物朝向城市的中央商务区,也有利于大楼顶端的植物园。

在这两座大楼中,较高的一座完全用于住宅,其高度为356.2m。这座大楼的顶部将设置一个对公众开放的花园。另外一座较低的大楼将用作酒店和商业空间,其最高点达到252.2m。

该建筑本身完全融入了墨尔本目前正在出售的文体娱乐和商业场所之中,然而其功能和连通性又更具多样性,因此设计进一步提供了一座混合用途的建筑,它本身就是一座城市。

垂直阶梯式的公共基础设施区域,囊括了娱乐、零售、办公、住宅、酒店以及展览等功能。该区域的室内外空间完美地结合了自然环境、公共空间和文化。

"城市绿脊"沿着墨尔本南岸区大道的方向向上延伸,将建筑的各种功能、文化、景观和可持续性等元素组织整合,使之浑然一体。"绿色脊柱"除了提供了各种便利设施之外,还带来了许多功能。

在建筑的地面层,"城市绿脊"毗邻南岸区大道,吸引人们来到这栋建筑,从"绿脊"处进入大楼,并沿着"绿脊"往上走,因此,"绿脊"在垂直方向上起到了拓展城市公共区域的作用。"绿脊"接着从裙楼顶层的公园开始,沿着两座大楼蜿蜒而上,最终停在住宅楼顶部的"未来花园"处。

The design proposal by UNStudio with Cox Architecture has been selected as the winning design, a more than $2 billion mixed-use tower in Southbank which will be the tallest in Australia, located in the heart of Melbourne.

UNStudio's design proposal aims to establish a new destination for the Southbank area and Melbourne. The proposal is integrally organised by one Big Detail: a "Green Spine" of vertically networked platforms, terraces and verandas. This multifaceted spine is created by splitting the potential single mass at its core, thereby forming two separate high-rise structures and causing them to reveal the almost geological strata of their core layers as they rise above a light-filled canyon.

As a result of this design intervention, the towers that emerge on either side can enjoy porous city views and

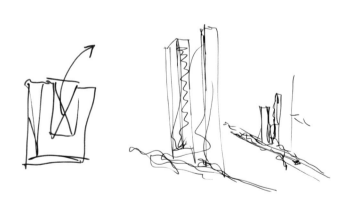

墨尔本网络扩展
Melbourne network extended

主要街道扩展
main street extended

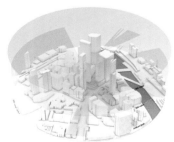

视野分析
view analysis

开放视野的分割
split to open views

太阳影响最大化
maximising sun impact

将绿意与城市连接
connecting green to city

露台
terracing

vastly improved contextual links, while the residences, offices and the hotel benefit from increased daylight and access to outdoor spaces.

The orientation of the Green Spine further enables an extension of the public realm on the podium, the continuation of green onto the towers and facilitates orientation to the CBD and the Botanical Garden at the top of the towers. The taller of the two towers will be entirely residential and reach a height of 356.2 meters. This tower will house a publicly accessible garden at its top. The lower tower will be home to a hotel and commercial space and top out at 252.2 meters. In addition to being fully integrated within the existing Melbourne network of cultural, entertainment, leisure and commercial venues on offer, with various programmes and its connectivities, the design further proposes a mixed-use building that is a city in itself.

A host of programmes, including recreation, retail, offices, residential, hotel and exhibition spaces are integrated into the vertically stepped public infrastructure – an infrastructure that is formed by indoor-outdoor spatial frames that embed nature, public space and culture.

The spine extends the Southbank Boulevard upwards and acts as the key organisational element of the building with respect to programme, culture, landscape and sustainability. In addition to housing a variety of amenities, all programmes are linked to the Green Spine.

At ground level, this spine directly engages with Southbank Boulevard by bringing people up and into the building, thereby expanding the public realm vertically. From the public park at the top of the podium, the Spine continues to entwine itself around the two towers, where it culminates at the top of the residential tower in "Future Gardens".

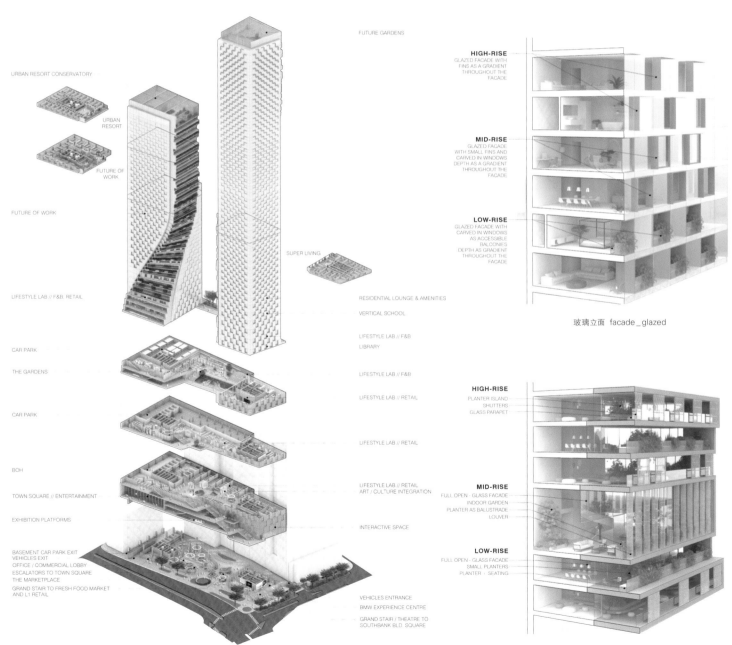

玻璃立面 facade_glazed

绿脊立面 facade_spine

景观中的智利住宅

Chilean Houses in the Landscape

在过去25年里，智利新兴产业的发展促进了其建筑风格的演变，因此该国家的建筑行业引起了国际上的关注。虽然这些建筑工作室共享着相同的地理位置与风光景色，但是每个工作室都在其领域内开发出了独一无二的设计方法。在接下来的几页内容中，我们将介绍最近由当代智利建筑事务所设计的五个项目，这五个建筑项目的选址均位于城市的边缘，展现了一国的建筑风格，旨在改善居民的生活，这些居民希望将自己从城市生活中解放出来，以享受当地自然景观的静谧。

During the last twenty-five years, Chilean architecture has drawn international attention due to architectural styles developed by emerging practices. These studios share the same geographic territory and landscape, but each of them developed approaches that are unique within their field. In the following pages, we will present five projects recently designed by contemporary Chilean offices, built at the fringes of the national territory as manifestations of an architectural style meant to enhance the inhabitants' life who seek to isolate themselves from the city life and enjoy the quietness of the local natural landscape.

Hualle住宅_Hualle House/Ampuero Yutronic
Loba住宅_Loba House/Pezo von Ellrichshausen
海滩别墅_Beach House/Schmidt Associates Architects
Ghat住宅_Ghat House/Max Núñez
H住宅_H House/Felipe Assadi Arquitectos
引人入胜的陈列室_Cabinet of Curiosities/José Luis Uribe Ortiz

引人入胜的陈列室
Cabinet of Curiosities

José Luis Uribe Ortiz

居住领域

几十年前,智利(反)诗人尼加诺·帕尔拉指出:在智利,"我们相信我们是一个国家,但事实是,我们只是一幅风景画"[1]。这一说法使我们能够就智利建筑的某些观点进行探讨,例如,建筑诞生于景观之中,建筑设计又界定了水平方位,这为建筑技巧与地域的发展设定了限制[2]。其他作者认为,智利的地理和景观在其建筑风格的发展中起着决定性的作用,该建筑风格侧重于将地域阐释为项目的起点,并注重最终的设计结果[3]。另一些人则认为,智利建筑的基本价值之一在于它能够以高度精细、多功能和抽象的方式与景观联系起来[4]。这些方法揭示了智利建筑的不同方面。在接下来展示的作品中可以认识它们,这些作品主要是关于第二套住宅或周末度假屋。这样的住宅可以看作是生活住宅,Juhani Pallasmaa是这样给它们下定义的:"家不仅仅是一个物体或建筑,而是一个分散而复杂的状态,它融合了记忆和图像、欲望、恐惧以及过去和现在。家是一个充满仪式感的地方,它可以包容个人节奏与日常工作。这些住宅是不能一次建成的,它们具有时间跨度和连续性;它们也需要有一个逐渐适应家庭和个体世界的过程。"[5]

Inhabiting the Territory

Several decades ago, the Chilean (anti) poet Nicanor Parra pointed out that: in Chile "we believe we are a country, but the truth is, we are just a landscape"[1]. This statement allows us to approach certain viewpoints regarding Chilean architecture, such as the fact that architecture is born in the landscape, and that architecture defines the horizontal orientation which sets limits between the artifice and the territory[2]. Other authors believe that Chile's geography and landscape have been decisive in the development of an architectural style which focuses on interpreting the territory as the starting point of the project and in the care given to the final result[3]. Others hold

这里精挑细选的作品都是一种建筑风格的体现，室内展示了居民的日常生活，他们试图将自己与城市的喧嚣隔离开来，并在宁静的景观中寻求庇护场所。这里的每一件作品都对应着不同的表现形式，这些表现形式探索着人们关于如何在这片土地上居住的想法。尽管它们使用与项目变量相似的相关条件，但是每一种建筑设计方法都是独一无二的。

碎片凝结

Pezo Von Ellrichshausen在他所著的《Intención Ingenua》[6]一书中，收集并叠加了一系列片段，使我们能够接触到建筑师处理其艺术和建筑作品的创造过程。这部作品承认了一种密度，这是建筑工作室持续生产的特征，从各种构图中可以看出已经撕裂过的纸张的重生过程、对类型学变量和油画颜料的探索（其中纯粹几何性的三维空间渗入到整个绘图之中），以及那些将自身视为片段、寻求彼此联系的元素。许多这样的碎片元素可以在Loba住宅（42页）中体现出来，该建筑位于科留莫曲折的海岸地形和崎岖的悬崖之间，远远俯瞰太平洋。这座建筑的特点在于它是一个孤立、封闭、抽象的建筑元素，是陆地与海洋之间的标志性建筑。Loba住宅是四四方方的，表面粗糙。而它的内墙却呈现出一种触觉体验，将粗糙的表面、渗透的混凝土和木质模具留下的痕迹真实地表现出来。圆形天窗贯穿整个建筑，可以将阳光引入建筑内部，并减轻建筑体量的重量。方方正正的墙壁将室内分为四个部分，阶梯式的地板区分了每个房间及其居住者使用它们的方式。这样的内部空间为居民提供了一个反思的机会，当他们向下穿过每一个楼层时，他们在移动的过程中会进行反思，最后到达一个可以俯瞰海洋的小平台，这样的构造最大限度地增强了孤立感和敬畏感。

居住的延伸

H住宅（94页）是由Felipe Assadi Arquitectos建筑师事务所设计的，这个项目通过将构成住宅的结构元素叠加在一起，从而创建出了

that one of the basic values of Chilean architecture lies in its ability to relate to the landscape in a highly elaborate, versatile and abstract way[4]. These approaches reveal different aspects of Chilean architecture. It is possible to recognize them within the works reviewed below which focus on second or weekend houses. These can be regarded as the living houses, which Juhani Pallasmaa defined as:

"The home is not simply an object or building, but a diffuse and complex state that integrates memories and images, desires, fears, and the past and present. The home is a setting for rituals, personal rhythms and day-to-day routines. Homes cannot be produced at one time. They have a temporal dimension and continuity; a gradual process of adaptation to the world of the family and the individual"[5].

The selection of works here presented is the manifestation of an architectural style in which the interior unfolds the daily life of an inhabitant who seeks to isolate him/herself from the intensity of the city, and take refuge under the tranquillity of the landscape. Each of these works corresponds to different manifestations that explore the idea of how to inhabit the territory. Although they assume similar contextual conditions as project variables, each architectural approach is unique.

Condensation of Fragments

The book "Intención Ingenua"[6] by Pezo Von Ellrichshausen, collects and superimposes a series of fragments that allows us to approach the creative process through which architects approach their works of art and architecture. This work acknowledges a density that is characteristic of the constant production of the architectural studio, which can be seen in compositions ranging from the reconstruction of torn paper, the exploration of typological variables and oils where the three-dimensionality of pure geometries saturates the painting, to elements that recognize themselves as fragments that seek to connect with one another. Many of these fragments are decanted in the Loba House (p.42), which is included between the sinuous coastal topography of Coliumo and rugged cliffs overlooking the Pacific Ocean. The house is characterised by a monolithic architectural element that is isolated, hermetic and abstract, and

一种空间对话，这样就提供了一个连接住宅与土地地形的入口平台。这种对话就介于自然结构与人造结构之间。虽然这个项目中的住宅起到了连接平台的作用，但游泳池却将自己与住宅分开，在风景中表现出来，就好像漂浮在附近的海面上一样。在《La casa de Zaratustra》一书中，Iñaki Ábalos指出："彻底的水平状态唤起了对神性本身的压制，也唤起了对任何垂直联系的抑制。它是对生活本身乐趣的一种表达，是对必须由房子来扩展的主题的一种近似。"[7] H住宅通过将住宅的中间部分膨胀成一系列外围护结构，优先考虑了居住在地平线上这一想法，创造了一个由面向大海的完全半透明的立面和面向山体的封闭性空间围成的界面。它通过在景观内提供暴露自然焦点的开口来给空间序列带来节奏与张力。每个外围护结构逐渐出现，给人们提供了不同的视野，并强调了住宅所投射出的距离。位于H住宅边缘的交通流线在建筑的一侧失去了流动性，并在那里形成了一个狭小的休憩空间，这个空间被两块平行于项目结构的混凝土板分隔。

折叠建筑
　　Felipe Grallert的文章《Barracas, tectónica y primitivismo》重点介绍了智利南部建筑的演变过程，并指出，"21世纪的某些当代作品看似遵循了1960年以前的原型，这些原型只能在大规模的生产或储存设施中找到。这些造型在建筑内部得到了重新设计，并衍生出了综合的理念，在内容和建筑之间建立新的关系"[8]。例如，从Ampuero Yutronic设计的Hualle住宅（26页）中可以看出Grallert描述的建筑风格。这个建筑作品最突出的地方在于它的海拔高度，以及它唤起维拉里卡火山形象的方式，其黑色玄武岩的外观和形状模仿了当地农村用于储存农业材料的建筑。在智利南部，建筑的复杂性是由资源的不稳定、现场建筑材料和业主的创造力（满足自身需要）所决定的。Hualle住宅在占主导地位的维拉里卡火山和当地乡村地形之间提供了一个狭小而有力的对话空间。建筑项目表现为一个中性的盒子，利用光的折射与反射原理，定义了连接其内部和外部的宜居空间。

provides a landmark between the land and the sea. The house is boxy, whilst its surface is rough. The interior walls of the Loba House suggest a tactile experience, manifested in a sincere way with crude surfaces, seeping concrete, and the marks left behind by wooden concrete forms. Perforating the building, circular skylights allow for sunlight to enter the building and remove weight from the building's architectural mass. Boxy stretches of wall divide up the interior into four volumes, playing with a stepped floor that differentiates each room and the way its inhabitants use them. This interior provides inhabitants with an opportunity for introspection as they move down through each level, ending with a small podium which overlooks the ocean and maximises a sense of isolation and awe.

Inhabiting the Extension
H House (p.94), designed by Felipe Assadi Arquitectos, generates a spatial dialogue by superimposing the structural elements that make up the home into a single collaborative piece that provides an access platform linking the house with the topography of the land. This dialogue mediates between the natural and artificial structures. While the house acts as a connection platform, the pool detaches itself from the house and juts out over the landscape as if it were floating above the nearby sea. In "La casa de Zaratustra", Iñaki Ábalos states that "Radical horizontality evokes the suppression of divinity itself, of any vertical link. It is an expression of the joy of life itself, an approximation of the subject that must be expanded by the house"[7]. H House prioritises this idea of inhabiting the horizon by distending a dwelling's intermediate level into a battery of enclosures, creating a perimeter articulated by a completely translucent facade facing the sea and hermetic features towards the hill. This gives rhythm and tension to the spatial sequence by providing openings to exposure natural focal points within the landscape. Each enclosure gradually emerges, providing different views and emphasising the distances projected by the house. The circulation at the edges of the H House loses its fluidity at one end of the house, where a small space of pause is formed, delimited by the two concrete slabs located in parallel to structure the project.

坡度测试

罗伯特·史密森在他的作品《沥青的倾泻》一书中记录了一辆自动倾斜卡车把沥青洒到碎石斜坡上的过程。在沥青倾泻下滑的过程中，我们可以在很短的时间内考虑侵蚀和沉积的自然过程。对于史密森来说，艺术和地理是研究地球上所发生的动态过程的一系列作品。与史密森一样，在Max Núñez的Ghat住宅（74页）中，正是地形清楚地阐明了作品的总体布局，作品从一个平行于斜坡的表面朝向了形成住宅主体的外围护结构。在将建筑物融入景观之前，设计师对土壤进行了调查，以了解先前存在的地形，作为设计结构的指南，从而将建筑设计成沿着山坡蜿蜒流入大海的形式。在设计项目中，房间在自然地层之间循环，自然地凸显了海洋和陆地的景观。由空心和实心空间混合构成的、高耸的空间序列构成了一个上升和下降的轨迹，这一空间由一系列柱子支撑，让居民以一种有趣的方式去体验美景。

中部空间

对于戈特弗里德·森佩尔来说，建筑的基本元素与火及其保存火的三种极限形式是相对应的。家庭用火倾向于分组使用，而空间界定则将其配置在灶炉内⁹。森佩尔提出的方法指导了Schmidt Associates Architects建筑师事务所设计的海滩别墅的总体布局（60页），该建筑试图通过将建筑的一部分融入到斜坡中来对居民进行分组规划。中央空间在私人围墙和位于房屋边缘的设施之间建立了一个共同点，以促进居民之间的互动，并在视觉效果和物理效果上使它们与海滩建立联系。该建筑设计通过适应斜坡的拓扑结构打破了其规律性。这座海滨别墅由两个位于基座之上的体量组成，基座调节了整个建筑群的密度和结构，试图将这座住宅展示成一个露台。

Folded Box

The article "Barracas, tectónica y primitivismo" by Felipe Grallert, focuses on the evolution of architecture in southern Chile and states that "Certain contemporary works from the 21st century seem to obey archetypes prior to 1960, only found in large productive or storage facilities. These shapes are reprogrammed in their interiors and derive in synthetic proposals that establish a new relationship between content and container"[8]. Hualle House (p.26) by Ampuero Yutronic could be seen an example of the style described by Grallert. The work stands out for its elevation and the way it evokes imagery of the Villarrica volcano with its black basalt exterior and shape which mimics the local rural buildings used for storing farm materials. In southern Chile; the complexity of architecture is conditioned by the precariousness of resources, the on-site building materials, and the owner's creativity to shape what is available into what he/she needs. The Hualle House provides a small, forceful dialogue between the large Villarrica volcano which dominates the landscape and the local rural terrain. The architectural project manifests itself as a neutral box which, using light folds and setbacks, defines the habitable spaces which connect its interior to the exterior.

Testing the Slope

In his work "Asphalt Rundown", Robert Smithson records the slippage of asphalt spilled by a dump truck onto a slope of rubble. As the asphalt descends the slope, it allows us to contemplate the natural process of erosion and sedimentation in a brief period of time. For Smithson, art and geography manifested as a series of works that investigated the dynamic processes which take place on Earth. Like Smithson, in Max Núñez's Ghat House (p.74), geography articulates the general layout of the work, which is oriented from a surface parallel to the slope into enclosures that generate the body of the house. Before the building was inserted into the landscape, the soil was surveyed for pre-existing topography and considered as a guide to design a structure which meandered down the hillside into the ocean. Rooms were designed to circulate amongst natural strata and emphasise views of the ocean and the land. Soaring spatial sequences, which are configured by empty and solid spaces, make up an

未来的智利

一方面，本节中介绍的项目有助于解释当前正在进行的辩论，即如今什么可以与当代的智利建筑相关联。另一方面，有人可能会争辩说，这些建筑可以被看作更为简朴的智利建筑，典型的周末聚会的极简化住宅。从学科的角度来说，使用与当地专业知识相关的语言进行试验的可能性，或者项目的形式，均可使用与现有环境相结合的一套当地设计原则来表达。这些考虑旨在为智利建筑景观的发展做出贡献，因为它们描述了当地生活的本质，同时也并没有低估该国未来将面临的不断变化的条件。

1. Nicanor Parra, 'Chile', *Obra Gruesa* (Santiago, Chile: Editorial Universitaria, 1973)
2. Miquel Adriá, 'Arquitecturas Chilenas', *Arquine: Revista Internacional de Arquitectura* (D.F. Mexico), No. 52, 2010
3. Fernando Pérez, 'La Excelencia en el Límite. Chile, Fortuna Crítica y Retos Renovados', *Atlas Arquitecturas del Siglo XXI: América* (Madrid, España: Fundación BBVA, 2010)
4. Josep María Montaner, Introducción al Suplemento 'Chile, Arquitectura Contemporánea', *Visions de l'Escola Tècnica Superior Magazine d'Arquitectura de Barcelona* (Barcelona, España), No 2, 2003
5. Juhani Pallasmaa, *Habitar* (Barcelona, España: Editorial Gustavo Gili, 2016)
6. Mauricio Pezo and Sofia Von Ellrichshausen, *Intención Ingenua* (Barcelona, España: Editorial Gustavo Gili, 2017)
7. Iñaki Abalos, *La Buena Vida: Visita Guiada a las Casas de la Modernidad* (Barcelona, España: Editorial Gustavo Gili, 2000)
8. Pedro Allonso, Umberto Bonomo, Macarena Cortés and Hugo Mondragón, *El Discurso de la Arquitectura Chilena Contemporánea* (Santiago, Chile: Ediciones ARQ, 2017)
9. VV.AA, *Escritos Fundamentales de Gottfried Semper* (Madrid, España: Fundación Caja de Arquitectos, 2015)

ascending and descending trajectory, supported by a system of columns which allows the inhabitants to discover the views appearing in a playful way.

Intermediate Space

For Gottfried Semper, the basic elements of architecture correspond to fire and the three forms of limits that preserve it. Domestic fire tends to group, while spatial delimitation configures it within a hearth[9]. The approach proposed by Semper guides the general layout of the Beach House (p.60) by Schmidt Associates Architects which seeks to group its inhabitants by inserting a part of the building into the slope. A central space establishes a point in common between the private enclosures and the facilities located at the edges of the house, promoting interaction among the inhabitants, as well as visually and physically connecting them to the beach. The architectural design breaks up its regularity by adapting to the topology of the slope. The Beach House consists of two volumes over a plinth that mediates between the density of the overall compound and its structure, in the attempt at showing the house as a terrace.

The Next Chile

On the one hand, the projects presented in this section contribute to the ongoing debates about what can be associated today with contemporary Chilean architecture. On the other hand, one might argue that these architectures can be seen as a more austere Chilean architecture, typical of a minimum dwelling in a weekend house. In disciplinary terms, the possibility of experimenting with language associated with local know-how, or as the form of the project, can be articulated to the formulation of a set of local design principles that merge with the existing environment. These considerations are intended as a contribution to the development of a new narrative about Chilean architectural landscape, for they describe the local essence of living without underestimating the continuously changing conditions to which the country will be subject in the future.

景观中的智利住宅 Chilean Houses in the Landscape

Hualle 住宅
Hualle House

Ampuero Yutronic

Hualle住宅位于智利的阿劳卡尼亚地区,该地区的自然景观盛名国内外。Hualle住宅是一座230m²的家庭住房。项目场地位于比亚里卡火山脚下一个坡度平缓的丘陵地带,俯瞰着下方的湖泊。该住宅坐拥周围的自然环境和地理位置带来的盛景。这座两层的建筑在形态和朝向上呼应了场地倾斜的地形和这个地区的微气候。

Hualle住宅坐落于这片土地的中心地带。从山路走近时就会发现,建筑在山坡上展现出来的形态会使人联想到这个地区的坡屋顶农用棚式建筑。相反,当从主山谷往北看时,会看到这座建筑融入了田园般的环境之中。

建筑形态

本案住宅与该地区的其他建筑(通常表面覆盖有黑色火山石)在外观上很相似,但实际上它采用的是在表面覆盖垂直的黑色软木板条的做法,软木板条上覆盖了一层黑色的防水织物,为建筑提供一种更为瞬时的皮肤般质感。从室内向室外的景观看,会发现窗户的大小和位置各不相同,这让建筑外观形成一种随机组合感。

室内墙面覆盖了颜色浅到几乎为白色的胶合板,与深色的外观形成鲜明对比。与此同时,窗户周围环绕着温暖的天然木材,精心地框住了组合的室外景观。天花板在室内的整个长度方向上呈现出蜿蜒的走势,其造型模拟的是屋顶结构的折叠几何形态。裸露的混凝土地板中加入了黑色火山石骨料,与浅色的墙壁和天花板形成了鲜明对比。

该建筑项目的最初设想是建造一个简单的矩形体量,但后来经过一系列调整和创新,建筑师创造出更加有雕塑感的形状:

1. 建筑向南延伸,形成不规则的楼层平面和体量;
2. 移除部分体量形成入口;
3. 一层北面的玻璃立面向内缩进;
4. 屋顶几何形状转变为一系列的折叠平面。

室内布局

该建筑的住宿区域紧紧围绕中央的双层高空间设置。这个中央空间设在南北轴线上,通过大型玻璃洞口可以充分欣赏到火山和湖泊的景观。北面的玻璃是一系列金属框的拉门,这些拉门利用了冬天太阳的低位,让阳光洒入空间内,而夏日的高阳则被外部悬挑顶遮挡。这个中央空间是家庭的社交核心区,一楼设有厨房、餐厅和起居室。中央空间的两侧,也就是建筑的东西两侧各设有一间客房和一间浴室。由楼梯走上二楼,经过工作室、图书室和一条俯瞰主中央空间的走廊就可以到达主卧和浴室。

可持续性/能源策略

由于地处偏远位置,这座住宅希望通过一系列的被动式设计和措施,最大限度地做到能源自给。
- 建筑朝向和高保温外墙(高于规范要求);
- 窗户的尺寸和位置根据所需太阳热量的多少而设计;
- 通过室内裸露混凝土表面的蓄热特点来调节室内温度;
- 供暖主要通过住宅中心的一个燃木火炉,这个火炉采用了独特的过滤系统,目的是最大限度地减少对外部环境的烟雾排放。

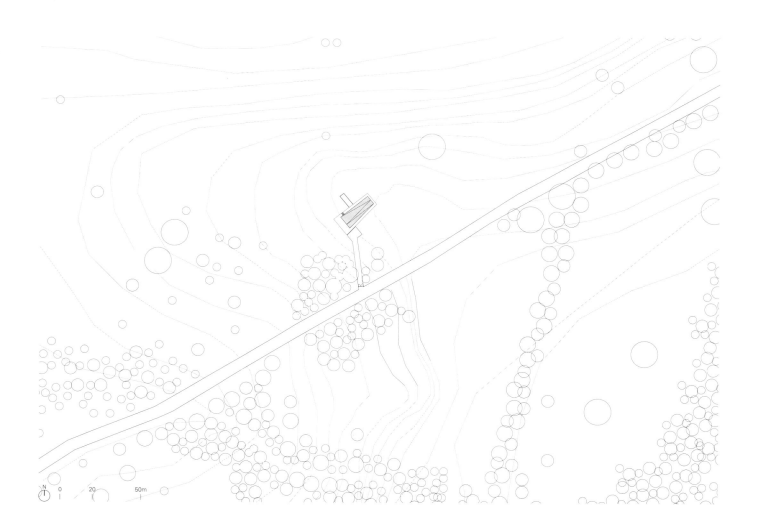

项目名称：Hualle House
地点：La Candelaria, Pucon, Chile
事务所：Ampuero Yutronic
项目团队：Javier Ampuero Ernst, Catalina Yutronic, Andy Wakefield
总承包商：Constructora Nuevo Horizonte
项目经理：Omar Loyola
结构工程师：Sigma
机械工程师：Clima Optimo
用地面积：5,000m²
建筑面积：160m²
总楼面面积：230m²
设计时间：2016
施工时间：2017.3—2018.2
摄影师：©Felipe Fontecilla

Hualle House is a 230-square-meter family home which is located in the southern Araucanía region of Chile, and this region is renowned for its natural beauty. Sitting on a gently sloping site in the rural foothills of the Villarrica volcano with a lake below, the house embraces this natural environment and the outstanding views afforded by its privileged setting. The two-storey building's form and orientation are a response to this sloping terrain of the land and the microclimate of the area.

The house sits on the central brow of the land. When approached from the road, the house is reminiscent of the region's agricultural pitched-roof sheds. In contrast, when

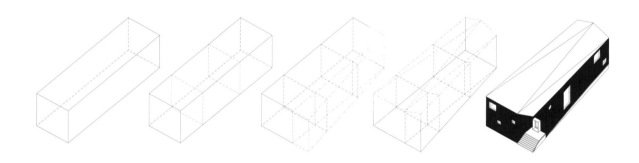

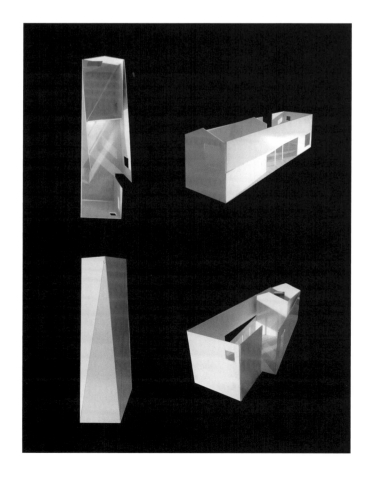

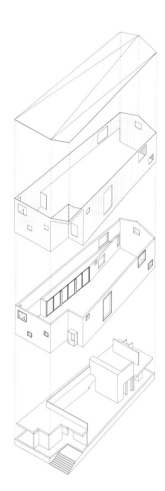

折叠屋顶 folding roof

黑色半条挡板 black slats screen

防水表皮 waterproofing skin

室内空间连接方式 internal articulation

viewed from the main valley to the north, the building blends into its pastoral context.

Building Form
The residence resembles other buildings in the area that are typically clad in the region's black volcanic stone. However, the dwelling is clad in vertical dark-stained timber slats, which have been placed over a black waterproof fabric that provides a more ephemeral skin-like quality. Dictated by the internal views towards the landscape, the windows vary in size and position, creating a seemingly random composition on the building exterior.

Contrasting the dark exterior, the interior walls are lined in pale, almost white, stained plywood. Meanwhile, warmer natural wood surrounds the windows, which frame carefully composed external views. Emulating the folding geometry of the roof structure, the ceiling flows sinuously for the entire length of the interior. Exposed concrete floors, incorporating black volcanic aggregate, contrast with the pale walls and ceiling.

Initially conceived as a simple rectangular volume, the design evolved through a series of manipulations and interventions to create the more sculptural form:
1. Extending the building footprint on the south side, creat-

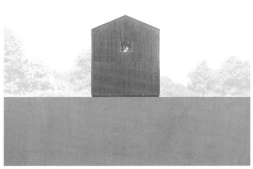

东北立面 north-east elevation

西北立面 north-west elevation

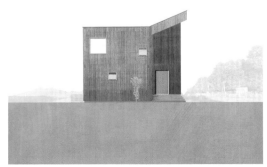

西南立面 south-west elevation

东南立面 south-east elevation

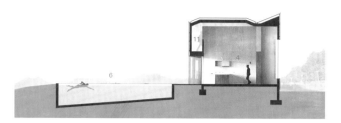

A-A' 剖面图 section A-A'

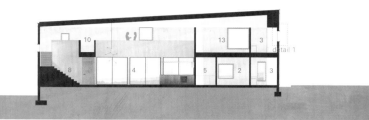

B-B' 剖面图 section B-B'

C-C' 剖面图 section C-C'

D-D' 剖面图 section D-D'

E-E' 剖面图 section E-E'

1. 入口大厅
2. 卧室
3. 卫生间
4. 起居室-厨房-餐厅
5. 餐具室
6. 游泳池
7. 烟囱
8. 楼梯
9. 学习室
10. 图书室
11. 走廊
12. 双层空间
13. 主卧

1. entrance hallway
2. bedroom
3. WC
4. living-kitchen-dining
5. pantry
6. swimming pool
7. chimney
8. stairs
9. study room
10. library
11. gallery
12. double height space
13. master bedroom

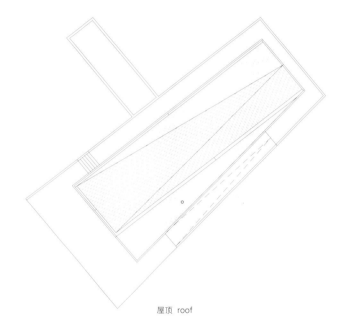

屋顶 roof

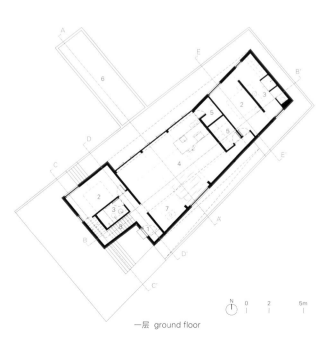

一层 ground floor

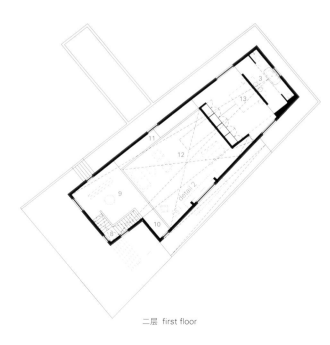

二层 first floor

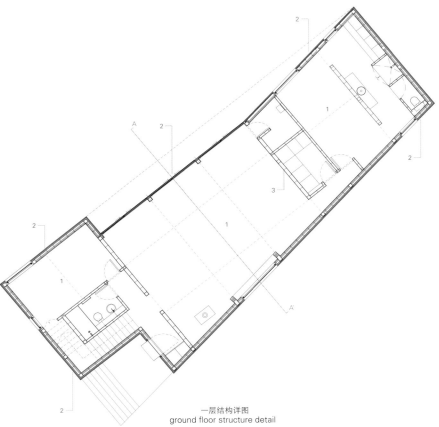

一层结构详图
ground floor structure detail

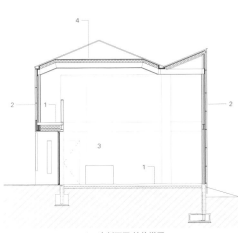

A-A' 剖面图 结构详图
section A-A' structure detail

1. floor build up
GF - exposed concrete floor with volcanic sand aggregates,
level 3 (semi-polished) / underfloor heating system / rigid insulation
damp proof membrane / hardcore soil ground
FF - exposed concrete floor with volcaninc sand aggregates,
level 3 (semi-polished) / metal decking / vapour control layer / non
combustible insulation / damp proof membrane / black painted
timber soffit

2. facade build up
vertical dark-stained 1"x 2" softwood slats every 10mm and fixed
with black screws, oil stained with 2 hands of Cutek and pigment
2X black ash / EPDM facade membrane / marine plywood / steel
structure / galvanised steel structural framing system / non-
combustible insulation / vapour control layer / premium paint finish
windows
Rauli wood frame / frameless double glazing bonded to black
aluminium window frame inset / black aluminium sliding double
glazing window system

3. internal walls
premium plywood / galvanised steel structural framing system
whitewash painted finish

4. roof build up
anthracite colour steel box profile sheets / EPDM roof membrane
marine plywood / steel structure / galvanised steel structural
framing system / non-combustible insulation / vapour control layer
plasterboard / matt white painted finish RAL9010

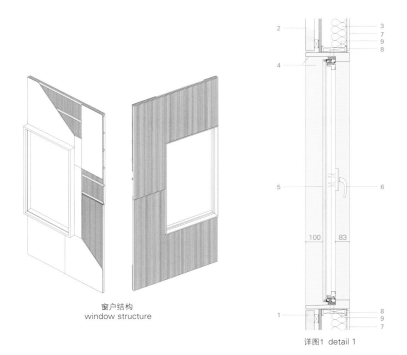

窗户结构
window structure

详图1 detail 1

window detail
1. vertical dark-stained 1"x 2" softwood slats every 10mm and fixed with
   black screws / oil stained with 2 hands of Cutek and pigment 2X black ash
2. EPDM facade membrane / marine plywood / steel structure
   galvanised steel structural framing system / vapour control layer
3. rockwool insulation
4. rauli wooden frame finish in natural colour
5. frameless double glazing bonded to black aluminium window frame
   inset / glazing 5/10/6 advantage clear cristal
6. black aluminium window handle
7. plywood premium 15mm grade B / white wash painted finish
8. galvanised steel structural framing system
9. wood frame reinforcement

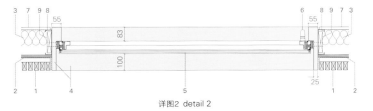

详图2 detail 2

ing an irregular floor plan and volume;
2. Cut-out in the volume to signify the entrance;
3. Recess of the north facing glazing at ground level;
4. Roof geometry transformed into a series of folding planes.

Interior Arrangement

The accommodation of the house is arranged simply around a central double height volume. This central space is orientated on a north-south axis with large glazed openings exploiting the views of the volcano and the lake. The north glazing, a series of metal framed sliding doors, takes advantage of the low winter sun entering deep into the space, which is otherwise protected during the summer months by the external overhang of the building. This central space is the social heart of the home, and accommodate the kitchen, as well as dining and living space at ground level. Two separate guest bedrooms and bathrooms are located at ground level to either side of this space on the east and the west sides of the house.

The master bedroom and bathroom on the first floor are accessed by stair via the studio, library and gallery walkway that overlook the main central space below.

Sustainability / Energy Strategy

Due to its remote location, the house seeks to be as environmentally self-sufficient as possible. This is achieved by a series of passive measures and interventions;
- building orientation and highly insulated exterior (beyond regulatory requirements);
- window sizes & positions that maximise or minimise solar gain where beneficial;
- utilising the thermal mass of the interior exposed concrete surfaces to regulate the internal temperature;
- heating primarily provided by a wood burning stove located in the heart of the house. This stove incorporates a unique filtration system which aims to minimise smoke emission to the external environment.

景观中的智利住宅 Chilean Houses in the Landscape

Loba 住宅
Loba House

Pezo von Ellrichshausen

也许物体和事物之间的唯一区别在于它们规模的大小。本案这座小建筑规模比较模糊，非常接近自然的事物，它更像是一个小屋，而非一座住宅：事实上它是一个小村舍。它是一座不透明的建筑，作为一个整体被紧紧固定在悬崖边，面朝着太平洋上的海狮保护区。这座建筑的厚度略显不足，造型既高又窄，就像是一面有人居住的、与自然地形垂直的高墙。这面墙的高度由两条线决定：一条是连续的地平线，另一条是阶梯序列，由六个面向大海逐级降低的平台组成。该建筑的水平屋顶（起到开放式露天平台作用）与地面规则的扩建空间（包括非正式的休息、用餐和起居空间）之间是分隔开的，起到分隔作用的是一个不对称的房间，房间内被三根巨大的柱子和两座连接桥打断。床铺被放置在上层平台上，那里的天花板较低，而沙发或桌子放置在垂直空间内较低的平台上。

建筑师在长长的建筑体量的两侧谨慎地开设了洞口，精准地设置了几扇天窗，此外还设置了几个可以用作日晷的半月形的洞口，以及一扇角窗——由圆形柱子隔开。这是唯一一扇与混凝土墙面齐平安装的无框玻璃窗。在落日余辉的照射下，一个几乎不可能的、看上去虚幻漂浮的岩石就出现在了反射光线中。

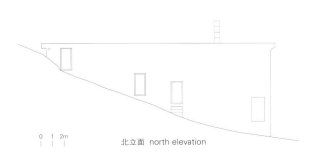
北立面 north elevation

西立面 west elevation

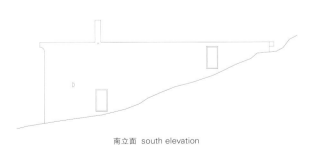
南立面 south elevation

项目名称：Loba House
地点：Coliumo Peninsula, Tome, VIII Region, Chile
事务所：Mauricio Pezo, Sofia von Ellrichshausen
合作者：Diego Perez, Thomas Sommerauer, Teresa Freire, Beatrice Pedroti, Wiktor Gago
建筑商：Carvajal & Cabrer
结构：Peter Dechent
顾问：Marcelo Valenzuela, Daniel Garrido
客户：Marcelo Sanchez, Janis Hananias
用地面积：1,000m² / 建筑面积：70m²
设计时间：2016 / 竣工时间：2017
摄影师：courtesy of the architect

Perhaps the only distinction between objects and things resides in their scale. Closer to any natural thing, in its ambiguous scale, this small building is more than a hut but less than a house: it is a cottage. As an opaque block, a monolithic object heavily anchored at the edge of a cliff, it is facing a sea-lion reserve on the Pacific Ocean. In its under dimensioned thickness, in its narrow and tall proportion, the building could be read as an inhabited wall that runs perpendicular to the natural topography. The height of this wall is deter-

mined by two lines: a continuous horizon and a stepped sequence of six platforms that descend towards the sea. The separation between the horizontal roof (with the function of an open terrace) and the regular extension of the ground (with the informal arrangement of rest, dining and living), is a single asymmetrical room, interrupted by three massive columns and two bridges. While beds are placed in the upper platforms with low ceiling, sofas or tables are meant to be in the lower platforms, within a vertical space.

There is a discreet regime of openings on either side of the long volume with some punctual skylights, a few half-moon perforations that could be used as sun clocks, and a singular corner window divided by a round pillar. This is the only window with unframed glass flushed to the outer concrete surface. Mirroring the sunset, an almost impossible and illusory floating rock rests right on top of that reflection.

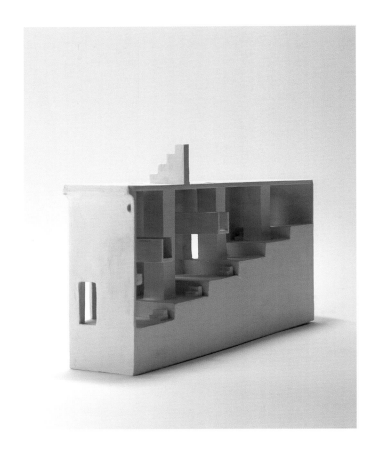

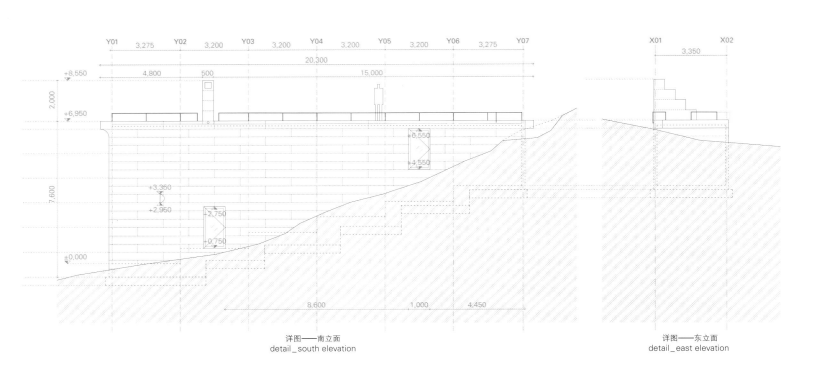

详图——南立面
detail_south elevation

详图——东立面
detail_east elevation

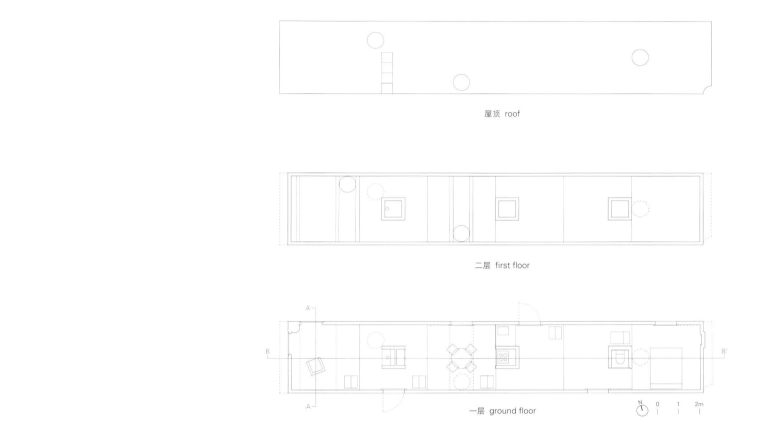

屋顶 roof

二层 first floor

一层 ground floor

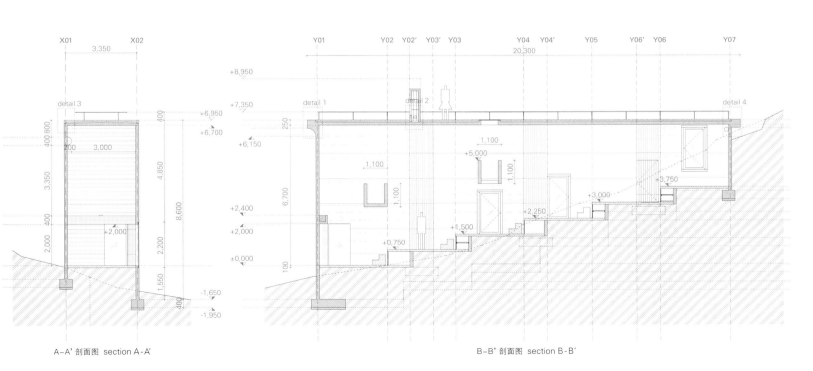

A-A' 剖面图 section A-A'

B-B' 剖面图 section B-B'

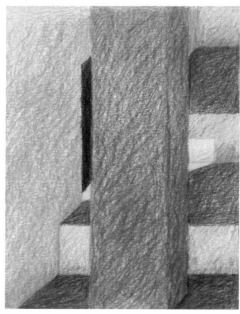

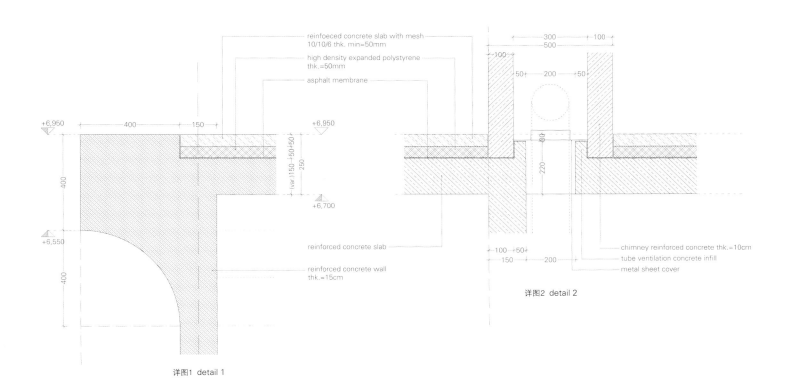

详图1 detail 1

详图2 detail 2

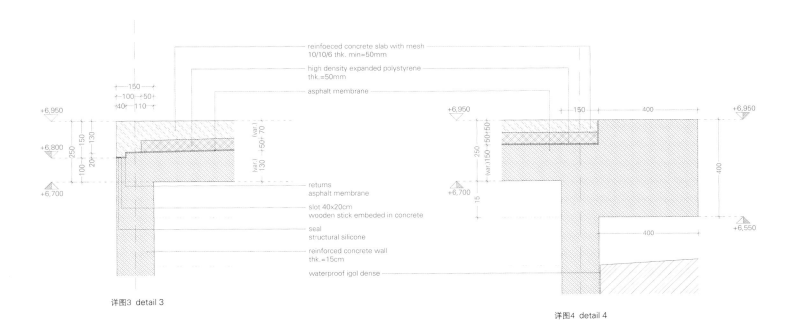

详图3 detail 3

详图4 detail 4

海滩别墅
Beach House

Schmidt Associates Architects

一条缓坡上长满了一层厚厚的波布罗特龙舌兰（又名海无花果），这样的景观是智利中部海岸的典型特征。此处虽然可以欣赏到壮阔的大海与美丽的日落，但却常常暴露在凛冽的西南风之中。加固过的沙坡共同支撑起宽敞的室内空间，该项目的室内空间由纯粹、简洁的材料建成。其成角度的几何形状充分利用了周围环境的景观。

一对大学教师夫妇和他们已经成年的孩子，希望能够拥有一座传统的海滩别墅以及这样的生活方式。别墅应该包含一间起居室、一间室内餐厅、一间可以直接从餐厅进入的厨房、一个能免受风吹日晒的室外用餐区、一个热水浴缸以及一间主卧室。他们希望从所有房间都能看到日落与远处的大海。海滩别墅的设计拥有更大的规划自由，空间整合得更好，从而打破了城市住宅传统的呆板和拘谨。

"海滩"是这座住宅的室外核心区,一切都发生在那里。海滩露台没有铺设坚硬的地板,而是仅用沙子覆盖,与放松休闲的理念相符,居住者可以赤脚漫步在海滩露台之上,享受回归淳朴的感觉。小巧的热水浴缸可供全年使用,一把形如石块的混凝土座椅陪伴在侧。这一适合沉思的地方是别墅主体量的一部分,与公共区域相连。

木质地板是该别墅唯一的硬质材料,铺设在入口处开放的走廊上以及儿童房与客房之间的连接区域,创造出了一个很大的无限公共区域。自然景观与所有建筑元素的结合是对远处海岸的重新阐释。

建筑形式的排列方式创造出了一个封闭的区域。主建筑高于地面且面朝北方,可以保护"海滩"区域免受风的影响。海滩别墅是一个独立的结构,包括起居室、餐厅、厨房、下方的服务区以及上方更加私密的主卧室。容纳儿童房与客房的东翼是完全独立的,它通过室外走廊与建筑主体相连,表明"海滩"区域始终是所有交流、聚会、事件和活动的中心。

该别墅是一座典型的钢筋混凝土住宅。它类似于土坯建筑,混凝土板的表面留有模具自身的纹理,并且全天均可以吸收日光,该建筑使用的材料既不复杂,维护起来也没有难度,因此减少了住户不必要的担忧。简单的建筑材料更能给人带来平和宁静的体验。为了弥补混凝土墙与地板冰冷质感的不足,建筑师设计了由层压木梁组成的屋顶和胶合板构成的天花板,这样的设计既能改善声学效果,表现出自然的纹理,又创造出了自然舒适的人性化环境。室内的装饰以及家具均采用天然材料制成,有助于营造令大家都满意的居住环境。

A gentle slope is covered with a blanket of Carpobrotus aequilaterus (a.k.a. sea fig), which is the characteristic of the central coast of Chile. While having a great view of the sea and the sunsets, it is exposed to the cold southwest wind. The consolidated sand slope joins to support the spacious interior of pure and simple materials of the project. Its angled geometry makes the most of the surrounding landscape.

An academic couple with grown-up children sought for a traditional beach house and such a lifestyle. The house should contain a living room, an interior dining room, a kitchen directly accessible from the dining room, and an external eating area protected from wind and direct sunlight, a hot tub, and a master bedroom. They desired to have a view of the sunset and distant sea from all the rooms. Beach House is designed to hold greater programmatic freedom and integration of spaces, with the consequent break from the traditional rigid and formal city houses.

"The Beach" is the central outdoor space of the house where everything occurs and happens. The beach terrace is not installed with a hard floor, instead is covered with sand to

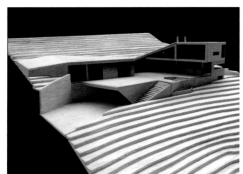

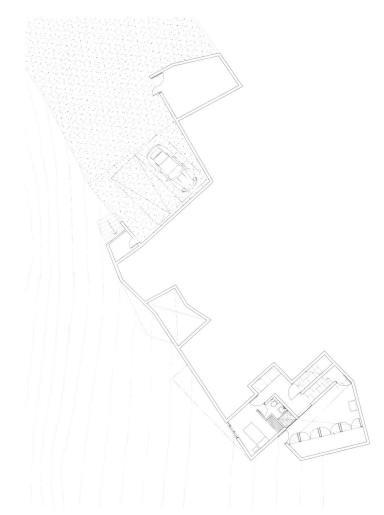

地下一层 first floor below ground

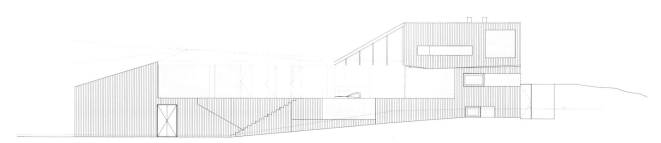

西立面 west elevation

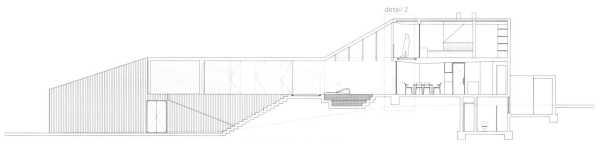

A-A' 剖面图 section A-A'

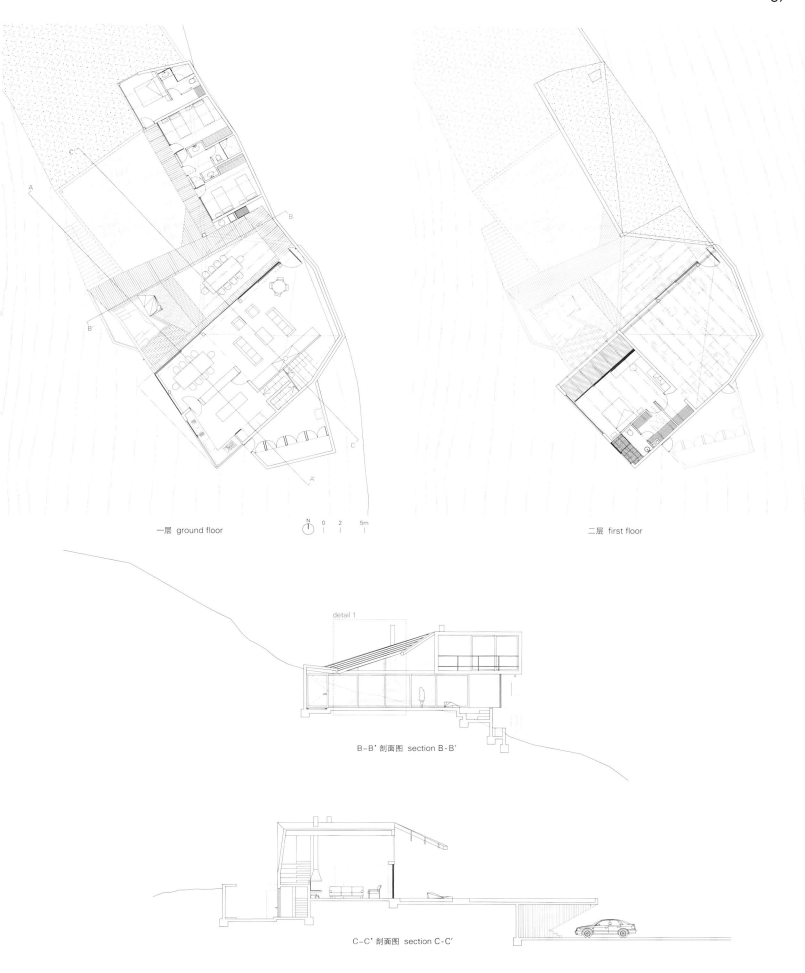

一层 ground floor

二层 first floor

B-B' 剖面图 section B-B'

C-C' 剖面图 section C-C'

项目名称：Beach House
地点：Beranda, Cachagua, V region, Chile
事务所：Schmidt Associates Architects
团队：Horacio Cortes Schmidt, Horacio Schmidt Radic, Martin Schmidt Radic
参与者：Nicolas Norero, Diego Ipinza
结构：Patricio Bertholet
建造商：C1 Construction
用地面积：4,200m² / 建筑面积：360m²
材料：wood, concrete, sand, glass, water
设计时间：2012 / 竣工时间：2016
摄影师：©Aryeh Kornfeld (courtesy of the architect) - p.60~61, p.62~63, p.64, p.65, p.68, p.69, p.70, p.72

support the idea of relaxation, so one can walk barefoot, without pretension. A small hot tub for all year round use is accompanied by a concrete deck chair that simulates a stone. This place of contemplation is integrated into the main volume of the house, connecting to the public areas. The only hard material, wooden decking, was used for the open corridor for access and for connecting the children's and guest rooms, generating a large common limitless area. The synthesis of a natural landscape and all built elements is a reinterpretation of the distant coast.

The arrangement of the forms creates an enclosed area. The

main block faces north and rises from the ground, protecting "The Beach" from the wind. It works as an autonomous body, incorporating the living room, dining room and kitchen, the service area beneath and the more private master bedroom above. The east wing that houses the children's and guest rooms is completely independent, while connected to the main body by an exterior corridor, showing that "The Beach" is always the center of all crossings, meetings, events, and activities.

The house was built as a typical reinforced concrete house. Resembling an adobe construction, the molds of the panels expose their texture and absorb the gleaming light throughout the day. No materials either sophisticated or difficult to maintain were proposed, thus reduce unnecessary worries; the simple materiality allows to enjoy the experience of peace and calmness. To compensate for the coldness of the concrete walls and floors, the roof is made of laminated wooden beams and the ceilings of plywood boards, improving the acoustics, delivering a natural texture, and humanizing the environment. Even the decoration and the furniture, made of natural materials, help to generate unprejudiced environment.

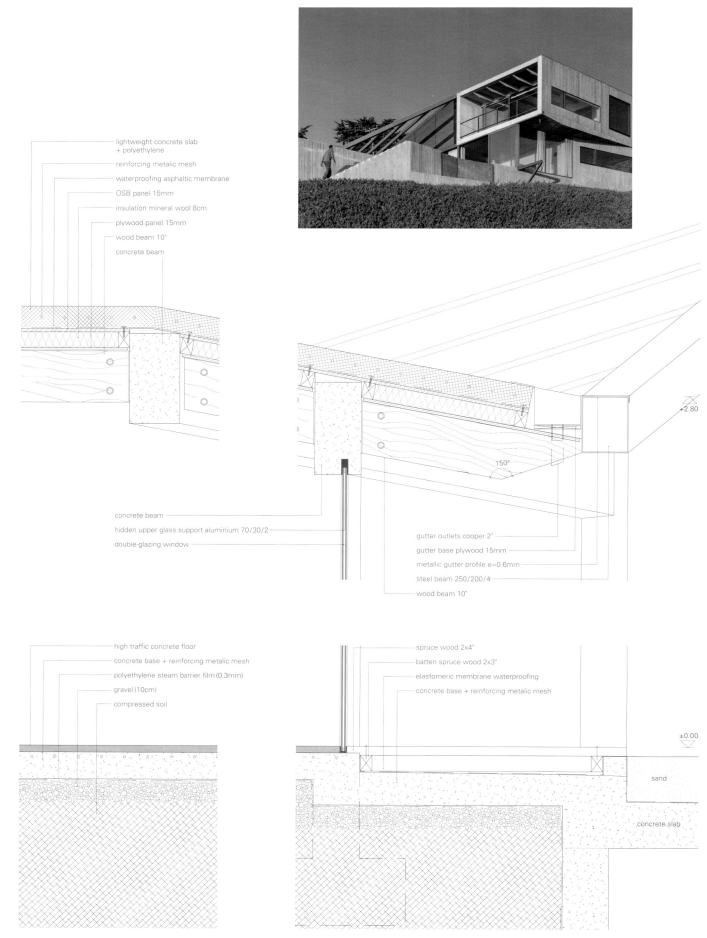

详图1 detail 1

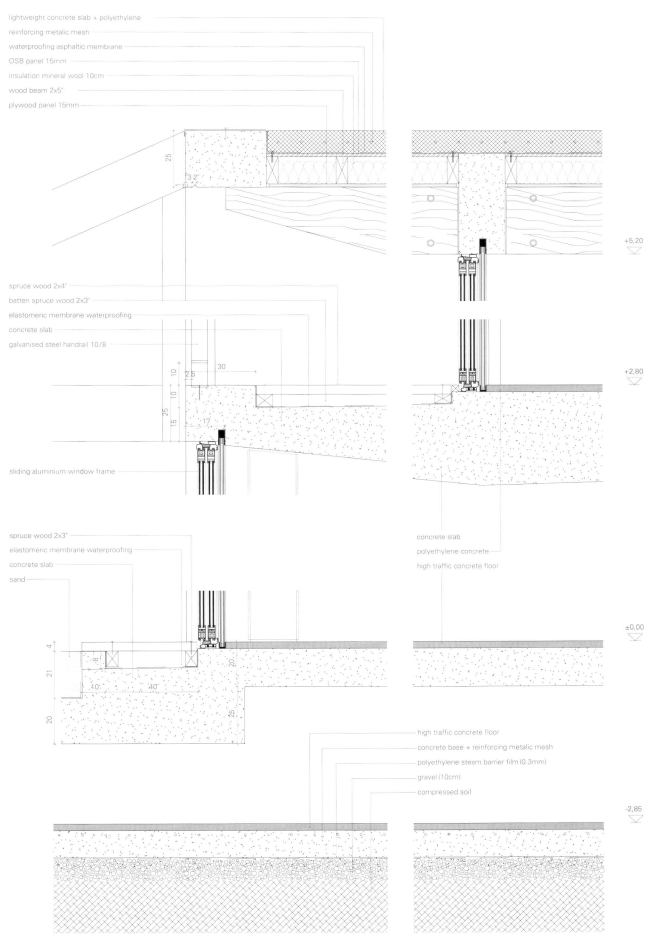

详图2 detail 2

Ghat 住宅
Ghat House

Max Núñez

Ghat住宅坐落于25°坡度的斜坡之上，面朝广阔的太平洋。其设计、结构、内部布局以及预期生活方式均取决于已有的地形。

屋顶的倾斜表面与场地的自然坡地平行，屋顶上设置了大面积台阶，这样的台阶规模在住宅项目中是十分罕见的。此斜面下的室内空间中包含了住宅的不同功能区。屋顶的斜面设计平衡了枯燥的自由平面，创造出有着丰富的尺度与高度的室内空间。

屋顶由15根形状、大小各异的混凝土柱子支撑。柱子的几何形态由该建筑的结构需求决定，其不同的形态设计将每根柱子个性化为独立的元素，避免了整座建筑形成死板的网格结构。每根柱子都在空间中形成一个独特的点，框定了周围不同的景观。

四个较轻的体量包裹了木质表皮，插在屋顶表面和下方空间之间。其中的三个体量是私密空间，第四个体量较小，直接从室内通向屋顶。这些体量分别位于屋顶的下方、旁边、上方，模糊了Ghat住宅的私密与公共空间之间的界线。

Ghat House is located on a terrain that is inclined 25° facing the Pacific Ocean. The design, structure, internal configuration, and lifestyle proposed within it were determined based on the pre-existing aspects of the topography. The inclined surface of the roof is parallel to the natural slope of the site and is covered with steps, unconventionally large for domestic use. The interior space under this tilted plane embraces the different programs of the house. The monotony of the free plan is balanced with the slopes,

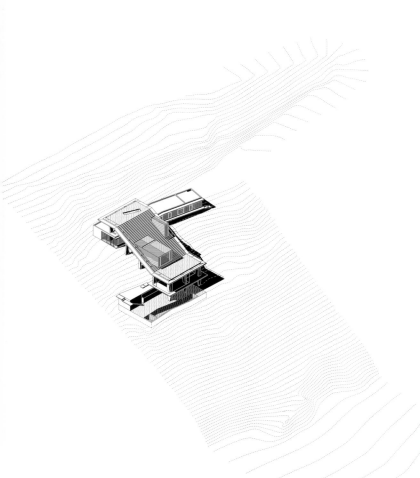

项目名称：Ghat House
地点：Cachagua, Zapallar, Chile
建筑师：Max Núñez
合作建筑师：Stefano Rolla
结构工程师：Mauricio Ahumada
建筑承包商：Francisco Álvarez
景观设计：Alejandra Marambio
照明设计：Estudio Par
技术检查：Alfonso Bravo
用地面积：2,297m² / 建筑面积：390m²
设计时间：2011—2014 / 施工时间：2013—2015
竣工时间：2015.5
摄影师：©Roland Halbe

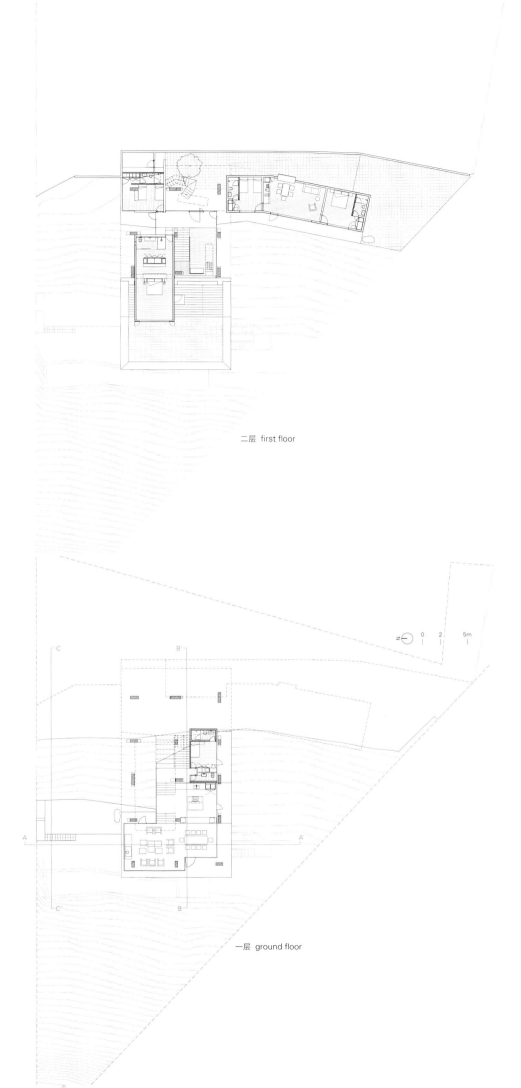

二层 first floor

一层 ground floor

creating an interior of varying levels with different sizes and heights.

The roof is supported by 15 concrete columns with different shapes and sizes. Their geometry was determined by the structural needs of the building and their heterogeneous shapes managed to individualize each column as a singular element, avoiding a rigid structural grid that dominates the layout. Each column forms a particular point in spaces, casting frames that yield various landscapes all around.

Four lighter volumes cladded in wood interfere with the surface of the roof and the space below. Three of these volumes contain private rooms and the fourth, smaller in dimension, includes direct access to the roof from the inside. These volumes, located under, aside, and above the roof, create obscure connections between the private and public areas of the house.

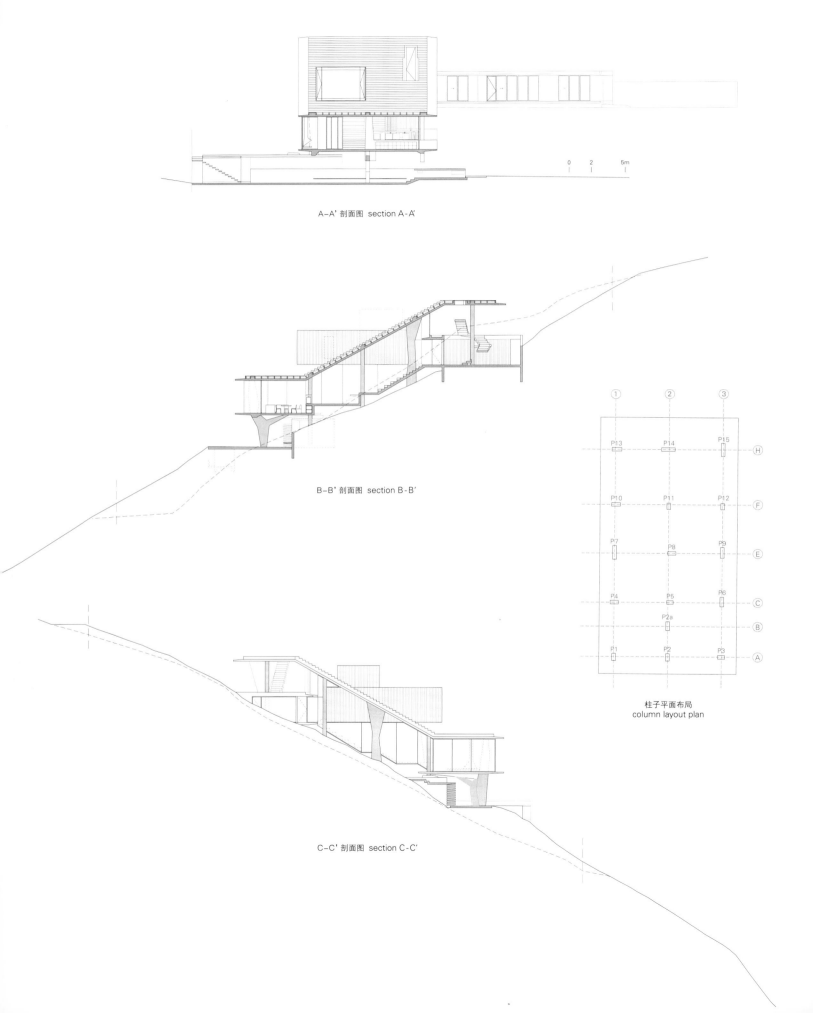

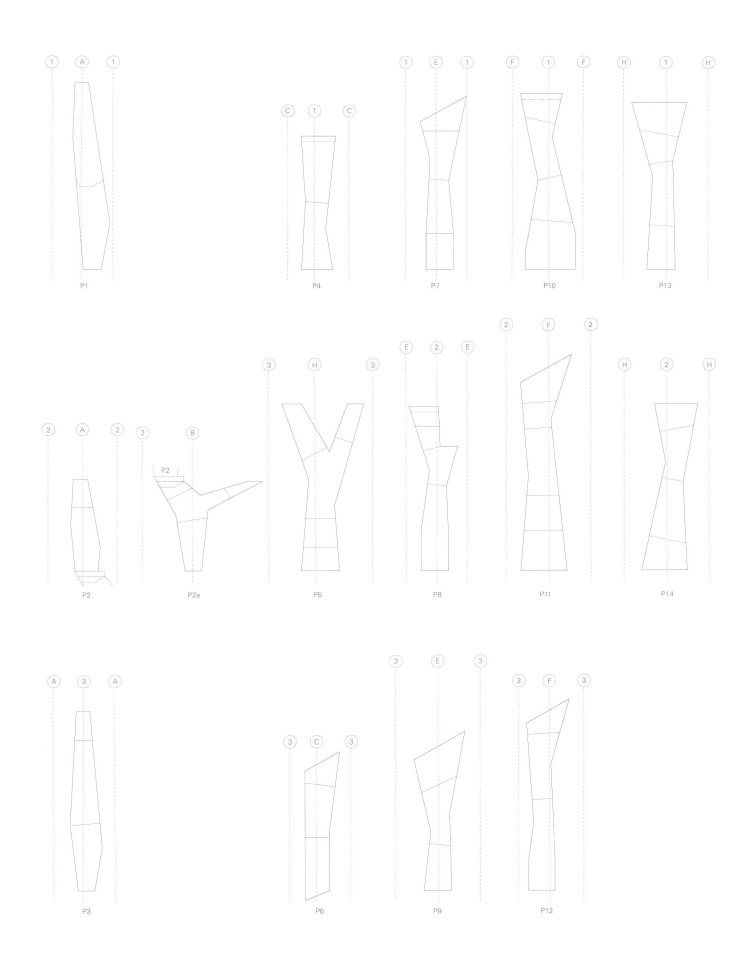

H 住宅
House H

Felipe Assadi Arquitectos

景观中的智利住宅 Chilean Houses in the Landscape

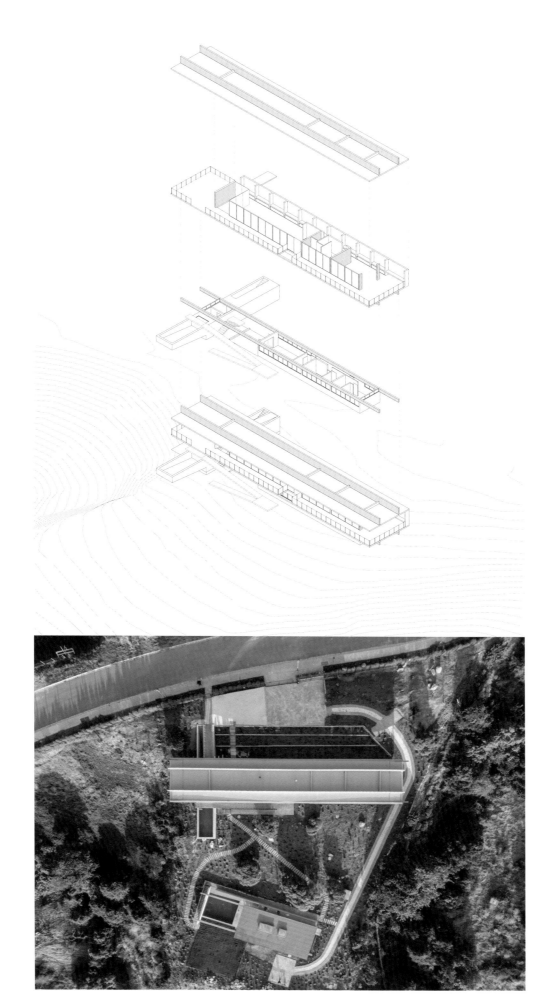

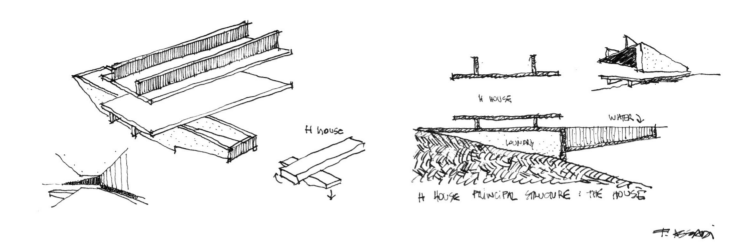

H住宅是一个钢筋混凝土结构，由一系列纵向和横向延伸的梁组成。这些梁共同作用，构成了一个整体结构。本项目的主题是居住在结构中，而不是构建一个房间，这个主题成为该项目中很重要的一部分。在成为住宅之前，此项目本身就是一个结构。

长41m的天花板悬吊在南北方向设置的两根主纵梁之上，纵梁高1.4m，仅由四面墙体支撑，在建筑的两端分别形成了一个悬挑出7m的大悬臂。该结构体系设在楼板下方另外两根长度相同的纵梁之上。相同的四面墙体维持着整座建筑的稳定，它们受力向下并延伸到楼板上，形成一个基座，这个基座的一侧是卧室，另一侧坐落在一个东西走向的横向建筑体量上。这个横向体量的一侧是住宅的入口，另一侧面朝大海且包含一个游泳池——一个截面有变化的厚重的楔形结构构成了另一个悬臂，面朝斜坡，同样也是悬挑出7m。

建筑的入口层设有公共区域，如起居室、餐厅和厨房，它们设在一个单独的空间内，没有隔墙或柱子分隔，这一层还包括一间主卧室和附属浴室。下面楼层设有室外通道，这一层包括一个家庭房以及次卧。巨大的梁墙构成了该住宅的主立面，这个梁墙是一件长长的木家具，贯穿住宅的整个长度，成为必要的围合结构。

一部楼梯和一条坡道组成了垂直方向的交通流线，它在两个楼层之间的连接处上升，在这个位置，横向的结构与泳池相会，楼梯和坡道通向庭院，该庭院自东向西穿过主体结构，加强了斜坡上悬浮的主体形象，斜坡也正是这个项目的灵感来源。

室内装饰尽量做到最少，以免分散人们的注意力。地面铺设着浅色的木质地板，天花板为裸露混凝土。涂了清漆的白色橱柜在厨房和浴室中显得尤为突出，反射着从室外照射进来的阳光。

House H is a reinforced concrete structure composed of a succession of longitudinal and transversal beams that work together to generate a single structural piece. The running theme of inhabiting a structure, rather than structuring a room, is very much a part of this project. Before being a house, the project was a structure in its own right. The 41-meter-long ceiling slab hangs from two main longitudinal beams set on a north-south orientation, measuring 1.4 meters high. The beams are supported by only four walls, forming large, 7-meter cantilevers at both ends. The system rests on two other longitudinal beams of the same length, located under the floor slab. The same four walls support the entire structure and project downwards to the floor, forming a base with bedrooms on one side while

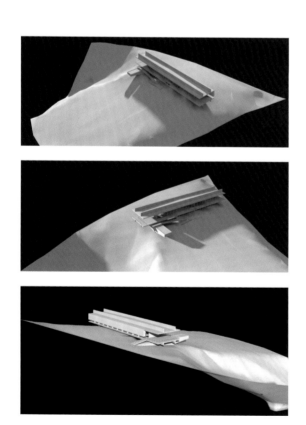

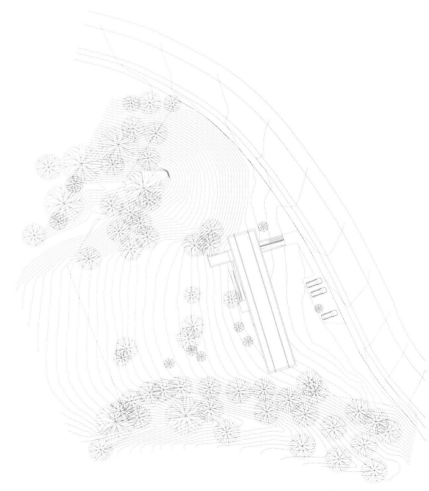

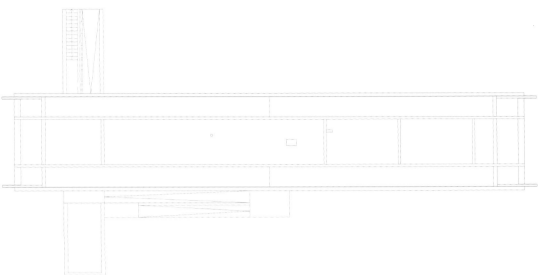

屋顶 roof

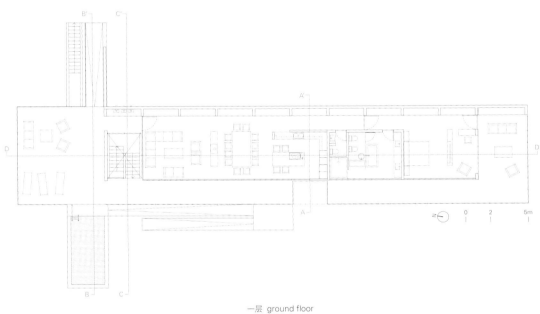

一层 ground floor

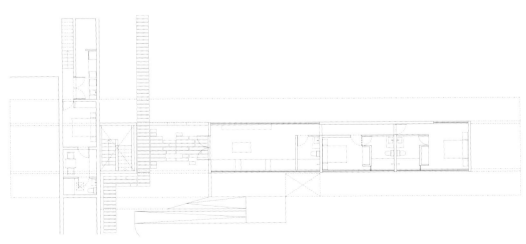

地下一层 first floor below ground

项目名称：House H
地点：Zapallar, Chile
事务所：Felipe Assadi Arquitectos
项目团队：Felipe Assadi, Trinidad Schönthaler, Macarena Ávila
Builder, General contractor: Alfredo Martínez
结构工程师：MP Ingenieros
土木工程师：Mario Pinto Maira
景观设计：Bernardita Del Corral
照明设计：LUXIA lighting
总楼面面积：340m²
竣工时间：2018
摄影师：©Fernando Alda

resting on an east-west transversal volume on the other. The latter volume frames the entrance to the house on one side, and on the other, projects towards the sea and contains the pool – a heavy wedge with a variable section that forms yet another cantilever, also 7 meters towards the slope.

The access level houses common areas such as the living room, the dining room, and the kitchen in a single space without partitions or columns, as well as the master

bedroom and the attached bathroom. The lower level, with outside access, contains a family room and the secondary bedrooms. The great beam-wall that frames the main facade of the house is a long piece of wood furniture that runs throughout the house, serving as necessary enclosure. A vertical circulation consisting of a staircase and a ramp rises at the connection point between the levels, where the transversal structure meets the pool volume. The staircase and ramp lead into a courtyard that crosses under the main structure from east to west, reinforcing the main imagery of levitation on the slope that inspired the project.

Interior decor is kept minimal, so as not to distract from the views. Light wood lines the floors, with exposed concrete on the ceilings. Lacquered white cabinetry features prominently in the kitchen and bathrooms, reflecting sunlight from outside.

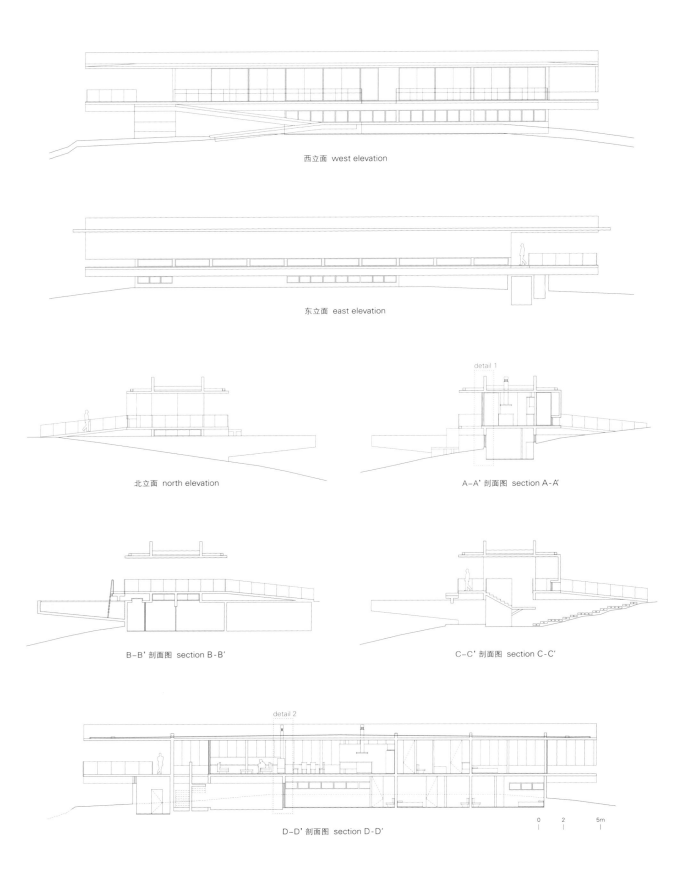

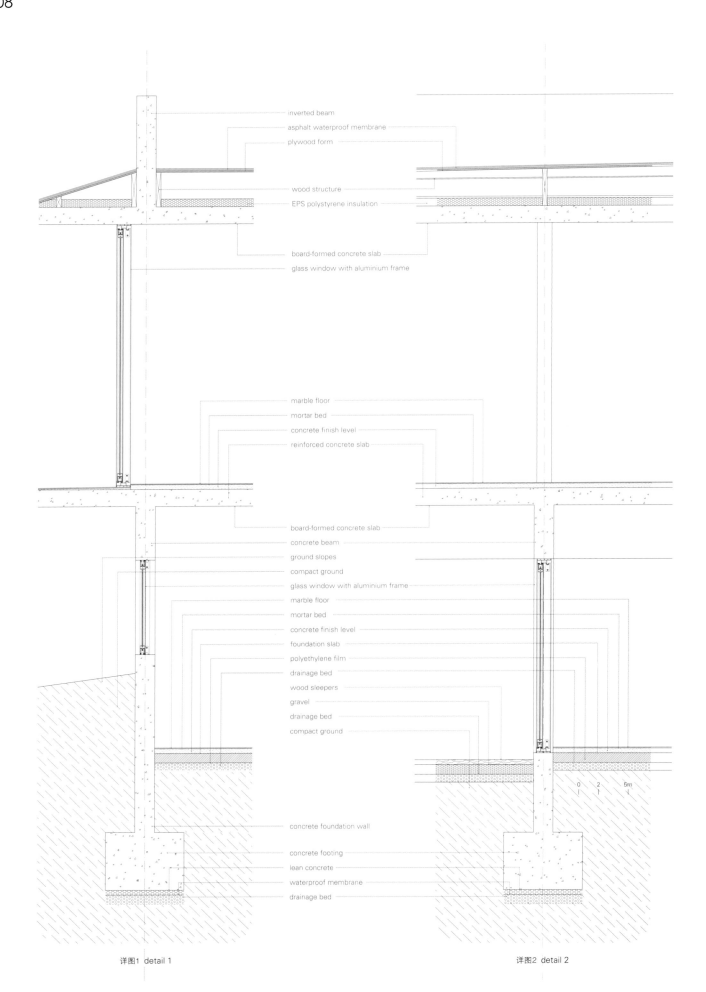

详图1 detail 1

详图2 detail 2

学生宿舍
非类型化建筑

Student Housing
Against th

有建筑评论家分析了某些以混合规划方案为特征的结构，这些混合规划方案体现了多种多样的功能和组合用途，他们将此类建筑定义为非类型化建筑，并称其是城市调和背景下社会和经济压力的产物。这类建筑已成为城市发展新进程的催化剂，新进程催生了不同寻常的聚集群落，与某些特定历史时刻发生的特定历史现象完美对应。近来，这种模式的重要性看起来似乎愈发明显。

例如，自21世纪初以来席卷中国的社会和经济转型就可以看作是这种超级城市形态出现的生发器，

Analyzing certain structures characterized by mixed programs that evinced a variety of functions and combined uses, some critics affirmed that such buildings, definable as anti-typological, were the product of social and economic pressures mediated by the city, which had become a catalyst for new processes that gave form to unusual aggregates perfectly corresponding to particular phenomena occurring in certain historical moments. Recently, this pattern seems to have assumed more and more importance.
The social and economic transformations that have swept China since the beginning of the 21st Century can be considered, for example, as generators of the Hyper-urban forms characterizing some recent architecture in cities such as Shenzhen, Beijing and Chengdu.

Olympe de Gouges学生宿舍_Olympe de Gouges Student Housing /ppa + Scalene + AFA
大田大学惠化寄宿学院_Daejeon University Hyehwa Residential College/Seung H-Sang
塔帕尔大学学生宿舍_Thapar University Student Accommodation/McCullough Mulvin Architects
城西大学爱家宿舍_Josai International University iHouse Dormitory/Studio Sumo
卢西安·考尼尔学生宿舍_Lucien Cornil Student Residence/A+Architecture

非类型化建筑_Against the Typology/Marco Atzori

这些超级形态构成了诸如深圳、北京和成都等一些城市中某些近期建筑的主要特征。而且，这些城市本身也正是这些新型实验性建筑类型的孵化器。
　　这种分析过程同样包括了主要住宅和临时住所（例如学生宿舍）的情况，其目的是研究因应居住者社会需求的变化而出现的规划性、功能性和社会性新型聚落的发展。如上所述，只有各种形式的非传统组合以及各种功能和用途的混合使用，才能对社会或其特定群体的当前所需做出特定回应。

Furthermore, it is the cities themselves that become the incubators of new, experimental architectural types.
This analytic process also includes the phenomenon of primary and temporary residence, such as student housing which, at this moment, is the object of research for the development of new forms of programmatic, functional and social aggregation in response to changes in the social needs of the inhabitants. As was stated above, only a heterodox combination of forms and a mixing of functions and uses can build specific responses to the current needs of society or its specific parts.

非类型化建筑
Against the Typology

Marco Atzori

  当前建筑研究中最有趣的一方面是建筑师能够设计出出人意料的规划性建筑组合形式,而非单纯地将一些建筑形式组合在一起,但这是他们所遵循的某种编码式设计语言或风格所拒绝的。
  通常,各种功能性项目组合在一起,外加一些主要功能的巧妙或补充性用途,就可构成正式建筑语言结构中的参考因素。这类建筑并非类型化建筑,而是与既定的参考因素分道扬镳,构成了新的建筑类型。
  当建筑使私人和集体空间之间产生关系、使个体维度与从属于某一群体的个体之间的关系也产生了联系时,这种创新就会伴随着更大的力量和兴趣诞生。集体住房的一个重要组成部分——学生宿舍,就是一个明显的例子。目前这方面的实验与20世纪提出的集体建筑有何区别? 最明显的区别不在于建筑语言,而在于集体空间的建构,这一建构形成了个性与集体关系的新概念。
  在现代运动或俄罗斯建构主义进行的"社会冷凝器"实验中,构建具有建筑师设想出来的或国家强加的生活维度的结构,这一意愿显而易见。相比之下,在当代建筑结构中,建筑师致力于寻求将用途、功能和时间相结合的可能性,意在激起居住者与建筑的惊喜邂逅(即使是人为控制的),而非以单一、确定的方式定义一系列的活动或生活方式。因此,建筑师更喜欢为意想不到的可能性创造机会,而不是坚持行为主义老一套或结构化的权宜妥协。此思想指导下的模式衍生出建筑物各部分之间的空间聚合。
  规划是一种形式-功能二项分解的空间序列生成器的内容在OMA、Steven Holl和Sanaa的建筑研究中有所表述,该研究对它在当

One of the most interesting aspects of current architectural research is the ability of architects to construct unexpected programmatic combinations rather than mere aggregations of forms declined by some coded language or style to which they adhere.
Often the combination of functional programs, as well as the addition of unexpected or complementary uses to the main functions, constitutes the reference parameters in the construction of formal languages. Such buildings do not flow from typological examples, but constitute new building types, often breaking with established references.
Such innovation happens with greater force and interest where the architecture gives shape to a relationship between private and collective spaces, between the individual dimension and the relationship among the individuals belonging to a community. One clear example is that important subset of collective housing: student dormitories.
What distinguishes current experiments along those lines from collective structures proposed during the 20th century? The most obvious difference is not in architectural language but in the construction of collective space that evinces a new conception of relationship between individuality and collectivity.
In the experiments regarding social condensers conducted by the Modern Movement or Russian Constructivism, the will to adhere to a construction of dimensions of living imagined by the architect or imposed by the state was evident. By contrast, in contemporary structures the architect, rather than defining in a univocal, deterministic way sequences of activities or lifestyles, looks for possible combinations between uses, functions, and times in an attempt to provoke unexpected events and experiences (even if controlled). The architect therefore prefers to create

Olympe de Gouges学生宿舍，法国
Olympe De Gouges Student Housing, France

代建筑语言发展中的重要性进行了清晰的阐述。我们可以视之为当代建筑理论建构的重要主题之一，并且我们还可以将项目概念的其他工具，如图表、自由剖面的概念以及室内空间的编码与之联系起来。

以下项目说明了规划的建构以及为集体活动而设计的空间的建构是如何影响建筑构想并最终使其具有特色的。

比如，ppa+Scalene+AFA设计的Olympe de Gouges学生宿舍（第120页）项目所制定的规划，就针对多个尺度，通过把功能与开放空间和闭合体量所构成的不间断序列联系在一起，进而将规模不同的校园、建筑物和宿舍联系到一起。

空间的灵活使用不会影响人们对建筑功能的清晰解读。"设计目标旨在创造一处活力四射而又舒适惬意的居住场所，以使住在这里的每个人都可以支配自己的空间。"

使用图表构建空间序列并将各种用途和功能组合起来，这一做法在Sumo工作室为爱家宿舍（第172页）制定的方案中显而易见，这也在寻求私人与集体空间的动态平衡中焕发了日本建筑文化中传统因素的活力。考虑到这些目标，建筑师们强调"面向走道设置多扇推拉玻璃门，能让人想到日本传统住宅中的'掾侧（连接传统和室和户外空间的木结构走廊）'空间。共享空间、走道和阳台系统将拥挤的生活空间拓展到了户外。"

在大田大学惠化寄宿学院（第136页）的设计中，IROJE建筑与规划师事务所的承孝相演示了可以在用途规划和功能规划之间生成

openings for unexpected possibilities rather than adhere to behavioral clichés or structured modus vivendi. The resulting matrices in turn generate spatial aggregations between the parts of the building.

The program as a generator of spatial sequences in which the form–function binomial breaks down is present in the architectural research of OMA, Steven Holl, and Sanaa, and in this work its importance in the development of the languages of contemporary architecture becomes clear. We can think of it as one of the great themes of the theoretical construction of current architecture, and to it we can connect other tools of project conception, such as the diagram, the conception of the free section, and the coding of the in-between space.

The following projects make evident how the construction of the program and of spaces destined for collective dynamics has influenced architectural conception to the point of characterizing it.

The project for the Olympe De Gouges University Student Housing of ppa + Scalene + AFA (p.120) develops, for example, a program that works on several scales and links the dimensions of the campus, buildings and housing through the connection of functions with a continuous sequence of open spaces and closed volumes.

The fluid use of the spaces does not compromise the clear reading of the functions. "Its ambition is to create an active and welcoming place to live in, where each resident can appropriate their own space."

The use of diagrams for the construction of spatial sequences and combinations of uses and functions is evident in the schemes produced by Studio Sumo for the iHouse Dormitory (p.172), which also evokes traditional aspects of Japanese architectural culture in the search for dynamics among private and collective spaces. With those aims in mind, the architects emphasize that "Multiple sliding glass doors open onto the walkways, recalling the 'engawa' space of traditional Japanese houses. The provision of shared spaces, the walkway, and balcony system expand

城西大学爱家宿舍，日本
Josai University iHouse Dormitory, Japan

动态关系的工具，构想出一个优先支持各种用户流的项目布局。

以上每座建筑都将各项功能视为一种城市机制而发挥作用，其中建筑与周边空间的关系至关重要，正如建筑本身通过经常将室内活动与周边景观联系起来进而实现室内外空间有效连通的能力也同等重要。在这一方面，最有效的工具之一是对诸如庭院等空间的运用：庭院允许外部环境在可控范围内接触建筑的周边，并保证了设定一项具有清晰功能的规划的可能性。

对这些项目的分析清晰表明了这些嵌入式庭院和开放空间的角色，以及建筑边界作为建筑和城市尺度中间协调者的作用，在这一方面会出现很多不同但却同样有趣的解读。

在A+Architecture设计的马赛卢西安·考尼尔学生宿舍项目中（第186页），大多数学生的房间朝向内庭院设置，建筑师将其定义为一个放松和共享空间，这一设计把学生宿舍和Saint Pierre街连接起来。这样一来，庭院就成为在城市尺度和建筑尺度之间起调节作用的城市元素。公共性质的活动可在内部广场上开展，一目了然，实现了庭院布局所提供的规划可能性。

即使是在Olympe De Gouges的大学宿舍里，在将活动从内部空间转移到外部空间这一层面，庭院也发挥着根本性的作用。在这种情况下，庭院成为各方关系体系的真正核心，它们允许建筑群落之间进行对话，并将建筑群落与周围的校园联系起来。

室内外关系的主题以及这些维度之间的界限（这也是外部世界的维度，亦是建筑中创造出的那片天地的维度），是承孝相在设计

the compressed living space into the outdoors".

In the Daejeon University Hyehwa Residential College (p.136), Seung H-Sang of IROJE architects and planners, demonstrates tools for generating dynamic relationships among programs of use and function, conceiving a project layout that prioritizes enabling a variety of user flows.

Each of the buildings contemplated herein functions as an urban mechanism in which the relationship with the surrounding space is of prime importance, as is the structure's ability to effectively connect the exterior and interior by continuously relating the activities that take place inside the building with the surrounding landscape. One of the most effective tools in that regard is the use of spaces such as courts that provide a controlled intrusion of the external environment into the perimeter of the building and allow the possibility of assigning a clear functional program.

The analysis of the projects makes clear the roles of these intruding courts and open spaces and of the quality of building limits as mediators between architectural and urban scales in allowing for quite different but equally interesting readings.

In the A+Architecture project for Lucien Cornil Student Residence in Marseille (p.186), most of the students' rooms look towards an inner courtyard, a place for relaxation and sharing, as defined by the architects, which connects the building with the rue Saint Pierre. The court thus becomes an urban element mediating between the urban and architectural scales. Events of a public nature clearly take place in the internal square, materializing the programmatic possibilities offered by the court's organization.

Even in the Olympe De Gouges University Student Housing, the courts play a fundamental role in the transfer

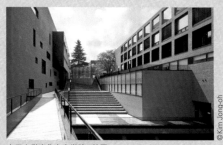
大田大学惠化寄宿学院,韩国
Daejeon University Hyehwa Residential College, Korea

惠化寄宿学院时所做各项考虑的基础。

由于寄宿学院被设想为一座修道院,因此它的围墙和通道作用也重要了起来,并披上了建筑形式的外衣。"修道院应该与外界有明确的界限,所以我建了一堵长长的墙,把学院和外界分隔开来。该建筑面向现有的校园区域开放,但由于地势偏僻,其内部构造从外部并不能完全看得到。"建筑的边界是通过巧妙地使用墙体和体量来打造的。在整个建筑体内,形式的聚集创造了大量的关系空间,并进一步与该地的地形和自然环境之间建立了联系。如此,建筑不仅成为了城市的一处界限,还同时变成了对自然的一窥之地,从而进一步清晰明确了其自身的作用。

在印度的塔帕尔大学学生宿舍建筑(第156页)中,McCullough Mulvin建筑师事务所把建筑与自然系统和周边景观尺度之间的对话,以非常强大的方式予以了放大。建筑物是作为景观的一部分而设计的,它们没有明确的角色和定位,但却加强了各种尺度之间的对抗和相互渗透。各建筑物之间保持着明显的同一性,但在形式上,在城市设计和内外连接体系方面,对建筑所运用的景观建构方法则极为清晰有力,以至于在建筑内部的空间组织方面也能清晰地反映出这一点。

私人空间方面呢? 如果说整个建筑物中的活动是对城市中各种动态的复制,那么起居空间之外的地方则相当于(城市中的)各种关系空间,而私人空间则是个人活动的舞台,是最小的空间单元——能满足个人基本需求的最小面积。在这种情况下,对其进行功能性

of activities from inside to outside buildings. In this case they become the real core of the system of relations between the parties, allowing a dialogue between the groups of buildings and putting them in relation with the campus that surrounds them.

The theme of the relationship between interior and exterior and the boundary between these dimensions (which are also the dimensions of the world outside and of the universe created in the building) are the basis of the considerations made by Seung, H-Sang in designing Hyehwa Residential College.

Since the building is conceived as a monastery, its limits and passages take on weight and an architectural form. "The monastery should have clear boundaries with the world, so I built up a long wall separating the college from the world. The architecture opens to the existing campus areas, but the inside of the academic monastery is not fully shown from the outside due to an isolated topography". The boundary lines are materialized through the skilful use of walls and volumes. The aggregation of forms creates spaces of relationship within the masses of the building and is further able to forge relationships with the topography of places and the natural environment. The building thus becomes a limit for the city and a pointing towards nature, thus recognizing its clear, distinct roles. The dialogue with the natural systems and the scale of the landscape is amplified in a very powerful way in the Buildings for Thapar University Student Accommodation(p.156) in India by McCullough Mulvin Architects. The buildings are designed as part of the landscape; they do not establish distinct roles and positions but enhance the confrontation and interpenetration of the scales. The architectural objects maintain their clear identity, but in the forms, in the urban design and in the connection system, both outside and inside, the construction of a landscape approach to architecture is extremely clear and powerful to the point to be reflected also in the internal organiza-

卢西安·考尼尔学生宿舍，法国
Lucien Cornil Student Residence, France

的组织是首要的：要以最大的关爱之心对最少的必备设施进行研究，在确保它们最佳完成使命的同时，在最小的空间中提供最佳舒适度。这种做法的隐含意思似乎是建议尽量减少对私人空间的使用时间，最好能尽快将之抛弃——这似乎是想说生命是外在的，社区成员之间的关系才应该是主流。从这种空间控制的角度来看，单元格式的私人空间设计不过是个有趣的挑战罢了，毕竟要使最小的空间发挥最大的功能，且同时保持绝对舒适度。窗户由此受到了极大的重视，因为它们除了能提供各种活动所需的自然光外，怎么说也是可以看向外部世界的地方，能够进而淡化个人空间的极端主义色彩，特别是在当下的数字时代。单元格都是自给自足的，因此在除了睡眠空间之外还提供必要的厨房设施。这些最小空间可满足学生们的主要活动需要，并在某些情况下保障他们对于个人极度隐私的选择。可以认为单元格意味着寻求最小生活空间的最低生活主义在当代背景下的逐渐式微。

　　正如建筑史中所常见的那样，这些集体建筑往往是拿材料和建筑技术做实验的地方，目的是换取有限成本条件下的质量保证。随着低技、低成本等概念以及责任型和资源意识型社会等假设的逐渐传播，新的美学标准已然确立，能够提倡并接受对于诸如工业或聚碳酸酯等材料的使用。A + Architecture将这些材料的运用作为卢西安·考尼尔学生宿舍的标志性规范之一，与此同时法国也出现了其他诸如Lacaton and Vassal所制定的实验性低技解决方案。同样的概念也被ppa + Scalene + AFA应用于他们所设计的Olympe De

tion of the building.
And what of the private space? When the building replicates the dynamics of the city, it transfers the spaces of relationship outside the living spaces, while the private space dramatizes the individual sphere and becomes a spatially minimal place: the minimum surface necessary to accommodate basic needs. In this case, then, the functional organization is primary: the minimum necessary structures and furnishing objects are studied with the utmost care and attention so that they fulfill their tasks in the best possible way while ensuring maximum comfort in the minimum space. The inference seems to be that the cell is to be occupied for the minimum time necessary, to be abandoned as soon as possible – as if to say that life is outside and that the relational dimension among the members of the community must be the prevailing one. From the point of view of space control, the cell is nevertheless an interesting design challenge, since the minimum space must attain maximum functionality but still be extremely comfortable. Windows thus acquire extreme importance because, in addition to providing the natural light necessary for activities, they are above all vistas onto the external world that can mediate the extremism of the individual sphere, especially in the digital age. Cells are almost always self-sufficient and therefore provide necessary kitchen equipment next to the sleeping space. These minimum spaces facilitate students' primary activities and guarantee them, in certain situations, the choice of extreme privacy. The cells can be thought of as contemporary declinations of the Existenz Minimum, the search for a minimal living space.
As has often happened in the history of architecture, such collective buildings are places for experimentation with materials and construction techniques to balance quality with cost containment. With the spread of low-tech and

塔帕尔大学学生宿舍，印度
Thapar University Student Accommodation, India

Gouges大学学生宿舍项目中，McCullough Mulvin 事务所在为塔帕尔大学学生宿舍选择材料时应用了工业材料的原因也是如此。

通过诸如减少不可再生资源的消耗、激活循环经济在材料的使用和回收过程中的作用等措施来控制生产过程中废气排放的需求，对建设过程和项目战略产生了深刻的影响。

从这一意义上来说，在这些建筑的建设过程中所使用的立面和材料，只是一种与外界进行能量交换的媒介，这种选择也逐渐将会是对保温、材料和外围护结构进行仔细研究和审慎选择的结果。在本文所分析的所有案例中，这种对技术性能的重视并没有减少建筑师们为保持建筑的高度复杂性而进行的研究。

毫无疑问，即使我们总会考虑到某一结构与建筑学科历史的关系，我们当今所处的不断变化的条件也依然在影响着建筑的观念，而且特征越来越明显。这个观念就是：建筑必须以有别于过往的模式来回应新的和不断变化的形势和结构的出现。过去的参考范例今天还依然有效吗？

low-cost concepts, including certain assumptions of a responsible and resource-conscious society, new aesthetic codes have become established, capable of proposing and accepting the use, for instance, of industrial or polycarbonate materials. A + Architecture makes the use of such materials one of the identification codes of the projects for the Lucien Cornil Student Residence, in line with other experiments in low-tech solutions conducted in France, for example, by Lacaton and Vassal. A similar concept is also applied by ppa + Scalene + AFA for their Olympe De Gouges University Student Housing, as well as the industrial matrix in the choice of materials for buildings made by McCullough Mulvin Architects for Thapar University.

The need to control atmospheric emissions in production processes, such as by reducing the consumption of non-renewable resources and activating processes of circular economics in the use and recovery of materials, profoundly affects construction processes and project strategies.

In this sense, the facades and materials used in the construction of these buildings serve as energy exchangers with the outside and, increasingly, are the result of careful studies and conscious choices concerning insulation, materials, and coating packages. However, this emphasis on technical performance does not prevent the architectural research from maintaining a high level of sophistication in all the cases analyzed.

It cannot be doubted that, even if one always takes into account a structure's relationship with the history of the architectural discipline, the changing conditions of our present are influencing the conception of architecture that, with ever greater specificity, must respond to new and changing situations and structures in ways that contrast with the modes of the past. Are models used as references in the past still valid today?

Olympe de Gouges 学生宿舍
Olympe de Gouges Student Housing

ppa + Scalene + AFA

开放的都市项目，活跃的校园

Ponsan-Bellevue大学校园的拆除和重建使得整个校园的氛围和形象得以重塑。新的学生宿舍位于校园边缘，旁边是一些小型的公寓楼和私宅。建筑的地址经过了仔细的选择，不仅考虑了目前当地居民对该地点的使用情况，还考虑了未来建筑的使用者们可能会如何使用它的问题。设计目标旨在为社区生活创造一处鼓舞人心的环境，一个活力四射而又舒适惬意的居住场所，以使住在这里的每个人都可以享受自己的空间。

该项目旨在将校园和邻近的Rangueil山连接起来，增强大学餐厅、大学技术研究院和周边社区之间现有城市的通达性。项目中心有一处较大的景观区，它将整个开发项目统一起来，延展了技术研究院的花园面积，并明确了校园的都市化规模。

该地点的地形便于在较高处建造学生宿舍，并且不会遮挡山丘。台阶从楼间延伸至一个较大的中央花园，花园继续向下，最后到达附近的技术研究院，它是整合和包围街区的"活动地带"，它们共同围绕着校园中心形成了一个同质化体系。

简朴、理性和节俭的设计，慷慨的建筑

新建造的宿舍腾出了一处公共地面来为学生们提供服务。这些设施为各种学生活动和社交生活提供了场地，并为滋养出活跃的校园社区氛围提供了条件。

建筑表皮上的镶板既是遮阳板，又是立面板，其抽象的排列方式回应了各建筑体量传统的修道院式布局。镶板的尺寸为整个项目平添了节奏感，改变了人们对项目规模的印象，并削弱了615间宿舍堆叠设计可能造成的蜂房印象。这层建筑表皮因周围环境的不断变化以及对其中居住者习惯的反映而一直在不断变化。表皮由覆盖系列金属镶板的木框架构成，金属镶板有从哑光到光滑共三种色调，排列组合在一起。框架的木制模块在车间预制完成，并在主框架完成后被立即直接固定在建筑的混凝土结构上。

对大型建筑而言，此种结构的选择可在有限的预算和时间内，提供高质量的组装和装饰效果。

根据地形和场地标高而精心设计的建筑方案，优先考虑的就是校园内的行人。因此，人们可以很容易地在同一连续地面上从一栋楼走

到另一栋楼。根据空间的不同类型、用途和所需专门营造的特殊氛围，设计师选择了丰富的材料和植物种类，并使其与Ponsan-Bellevue校园的整体景观相一致。

舒适灵活的宿舍空间

建筑物的上层分布着面积为16m²的标准宿舍空间，在是否要使其完全适应学生们的要求方面，并没有什么过多的规定。装修时仅使其满足了烹饪、工作、娱乐和休闲等主要需求，留出了更多起居空间，便于学生们按照个人用途和生活方式对空间进行重组。在技术层面之外，项目的目标是使设施在用途、美学和学生们不同生活方式、文化和性格的适应性方面都得以持久使用。

An Open, Urban Project, An Active Campus

The demolition and reconstruction of the Ponsan-Bellevue university campus encouraged the renewal of the atmosphere and image of the entire campus. The new student housing is situated on the edge of the campus, alongside small apartment buildings and private housing. Thus, the site was carefully treated in the way it is being used by local residents, as well as the way it might be used by its future occupants. The ambition was to create an encouraging setting for community life, an active and welcoming place to live in, where each resident can appropriate their own space.

This project aimed to link the campus and the neighbouring Rangueil hills, reinforcing the existing urban permeability between the university restaurant, the University Institute of Technology (IUT), and the surrounding neighbourhoods. A large landscaped area at the centre gives a unity to the entire development, extending the IUT gardens and defining the urban scale of the campus.

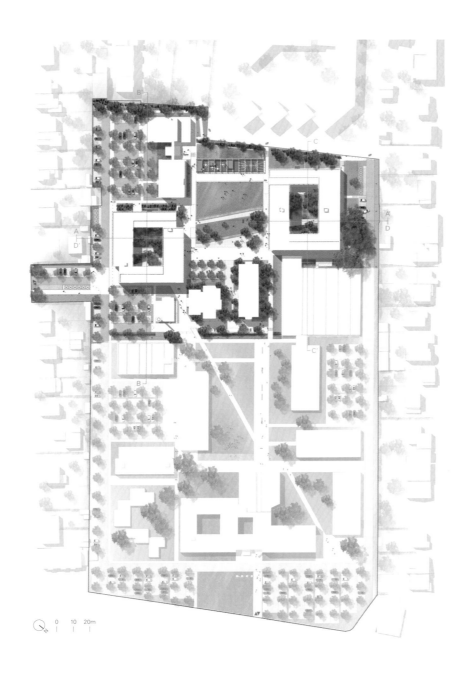

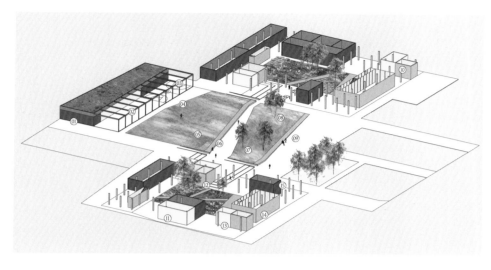

1.DIY	1. DIY
2.边喝边聊	2. discussing over a drink
3.打篮球	3. playing basketball
4.掷飞碟	4. playing frisbee
5.聚会	5. partying
6.坐着读杂志	6. sitting and reading a magazine
7.在草地上阅读	7. reading on the grass
8.小睡	8. taking a nap
9.会友	9. meeting with some friends
10.等电梯	10. waiting for the elevator
11.打听信息	11. asking for information
12.放下行李	12. dropping his luggages
13.取邮件	13. getting my mail
14.停自行车	14. parking his bike
15.和朋友们一起学习	15. studying with some friends

1. 高强螺栓
2. 焊接在转动铰链轴上的70×6扁钢
3. 25×25方轴
4. 支撑轴的6mm板
5. 闪银色外墙铝塑板覆层
6. 不锈钢开窗杆，直径10mm
7. 30/10不锈钢薄板滑轨
8. 50×30×2横管
9. 百叶联锁系统
10. 窗框和40×40×2横管
11. 门挡
12. 40×10扁扶手
13. 3mm厚铆接板
14. 窗框上的直径5.1mm预穿孔，与安装闪银色外墙铝塑板的孔对齐
15. 20/10外侧面板
16. 闪银色外墙铝塑板百叶覆层
17. 闪银色外墙铝塑板
18. 金属轨道型材
19. 金属盖板型材
20. 12.5mm费马策防火板
21. 50×120木壁骨
22. 120mm厚岩棉立面保温层
23. 18×50通风板条
24. 覆层节点处的褶层
25. 金属板
26. 50mm厚岩棉保温层
27. 隔汽层
28. 保温层
29. 双层丙烯酸丁酯18
30. 直径6.5mm L75螺栓
31. 岩棉保温层
32. 2×锚固件，直径12 L110
33. 混凝土板

1. high-resistance bolts
2. steel flat 70x6 welded onto the gudgeon of pivoting hinge
3. square gudgeon 25x25
4. 6mm thick plate supporting the gudgeon
5. alucobond cladding
6. stainless-steel opening rod, 10mm diameter
7. rail of stainless steel sheeting 30/10
8. horizontal tubing 50x30x2
9. shutter blocking system
10. opening frame and horizontal tubing 40x40x2
11. door stop
12. flat handle 40x10
13. 3mm th. riveted plate
14. 5.1mm diameter holes pre-pierced in the opening frame, aligned with the holes for the alucobond
15. lateral cladding sheet 20/10
16. alucobond panel shutter cladding
17. alucobond panel
18. profiled metal rail
19. profiled metal cladding
20. fermacell fire board 12.5mm
21. timber studding of 50x120
22. rockwool facade insulation, 120mm th.
23. ventilation lath 18x50
24. fold at junction between cladding
25. metal sheet
26. rockwool insulation 50mm th.
27. vapour barrier
28. insulation
29. 2xBA 18
30. screw diam 6.5mm L75
31. rockwool insulation
32. 2 anchoring pins diam 12 L110
33. concrete slab

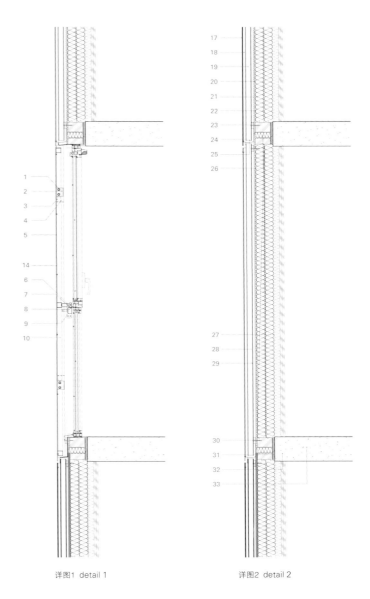

详图1 detail 1   详图2 detail 2

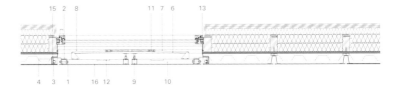

详图3-关闭 detail 3_closed

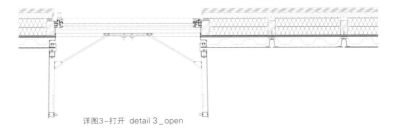

详图3-打开 detail 3_open

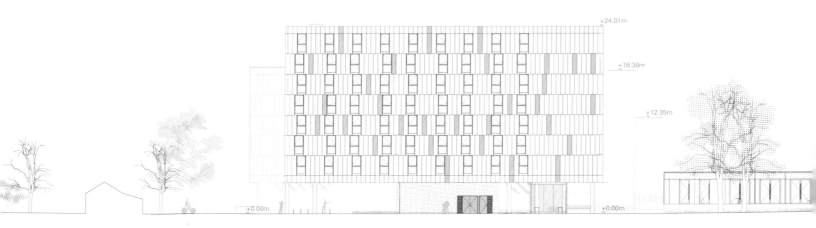

东北立面 north-east elevation

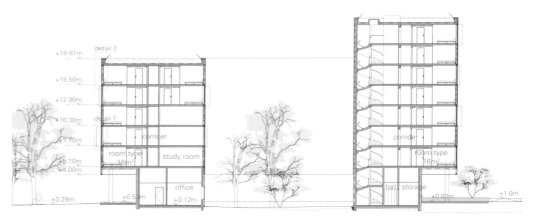

A-A' 剖面图 section A-A'

B-B' 剖面图 section B-B'

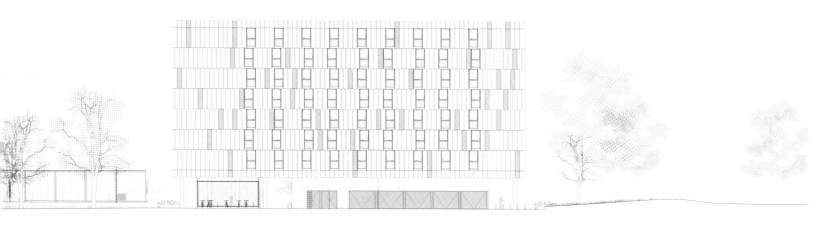

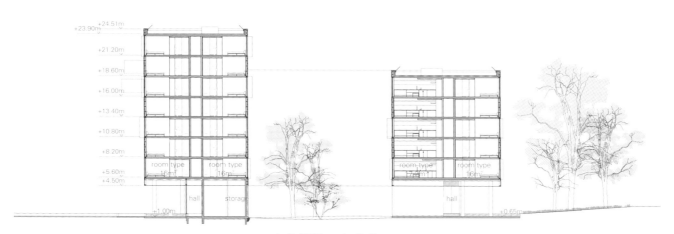

C-C' 剖面图 section C-C'

D-D' 剖面图 section D-D'

项目名称：Résidence universitaire Olympe de Gouges / 地点：35 rue Maurice Bécanne, 31000 Toulouse, France / 建筑设计：ppa architectures + Scalene architectes + Almudever Fabrique d'Architecture / 景观设计：ATP / 家具和宿舍设计：a+b / 标识设计：documents / 工程：Egis / 成本控制和现场监理：Execo / 业主：Nouveau Logis Méridional (Groupe SNI) / 用户：CROUS Toulouse Occitanie / 项目：demolition, reconstruction of 615 student lodgings, 3 staff apartments, communal living building (events hall and CROUS cultural services for students), landscaping / 总楼面面积：14,641m² / 成本：21,88 M € (including demolition and asbestos removing + roads and utilities and outdoor areas) / 竞赛时间：2013.9 / 建筑许可：2014.4 / 拆除：2014.11 / 施工时间：2015.2—2016.7–phase 1 (north island); 2016.6—2017.8–phase 2 (south island) / 完工时间：2016.7—phase 1; 2017.8—phase 2 / 开放时间：2017.10
摄影师：©Antoine Séguin (courtesy of the architects) - p.123, p.124, p.125, p.131upper, p.132~133, p.134, p.135right-lower; ©Philippe Ruault (courtesy of the architects) - p.120, p.122, p.128~129, p.130, p.131lower; ©a+b (courtesy of a+b) - p.135right-lower

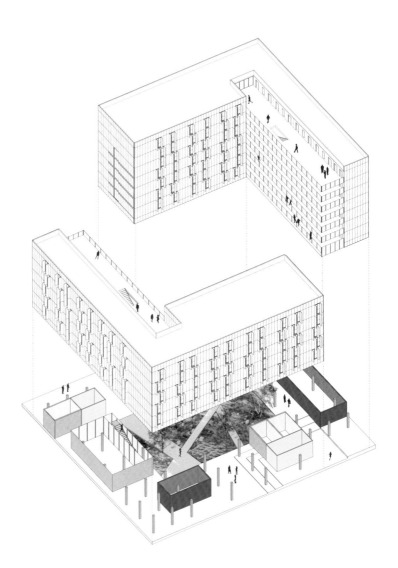
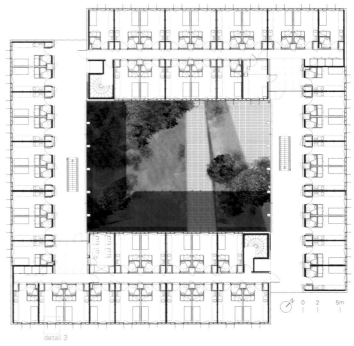

南侧建筑 south island

The topography of the site made it possible to arrange the student buildings on the higher ground without hiding the hills. The terracing extends out from the buildings through a large central garden that gradually descends down to the neighbouring IUT, which serves as the "active strip" that integrates and encompasses the neighbourhood, forming a homogenous ensemble around the campus centre.

Simple, Rational and Frugal Architecture, Generous Buildings

The buildings are lifted up to liberate a communal ground level for student services. These facilities provide a structure for student activities and social life, nurturing conditions for an active community campus.

To the ancient, cloister-like arrangement of the volumes, responds the abstraction of the panels, which serve as both shutters and cladding on the facade. Their dimensions add rhythms to the entire project, changing the perception of scale and diluting the beehive impression that might otherwise arise from the repetitive stacking of 615 lodgings. This outer skin always changes to react to its surroundings and reflect the habits of its occupants. The skin is formed of a timber frame clad in metal panels in a range of three different finishes, from matt to satin. The timber modules were prefabricated in the workshop, and then fixed directly to the building's concrete structure as soon as the main frame was sufficiently constructed.

This choice of construction comes with a high quality of assembly and finish, for very large-size objects, within a limited budget and timeframe.

Careful design with the topography and levels of the site prioritizes pedestrians within the campus. Consequently, residents can readily move from one building to another on a continuous level. A palette of materials and plants has been selected according to different kinds of spaces, their uses, and the particular atmosphere to be created, all consistent with the overall landscaping of the Ponsan–Bellevue campus.

Comfortable and Flexible Lodging

Distributed on the upper floors of the buildings are the lodging (16m^2), a standard space that has not been over-prescribed to remain freely adaptable to its occupant. It is furnished to fulfill the main requirements of cooking, working, entertaining, relaxing, etc, so as to enable reorganization for individual use and lifestyles, while freeing up living space. Beyond the technical aspects, the objective is the durability of the installation in terms of use, aesthetic, and its adaptability to the different lifestyles, cultures, and personalities of the students.

133

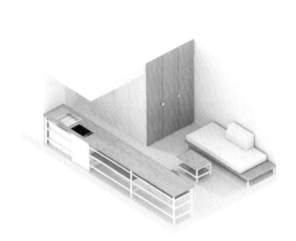

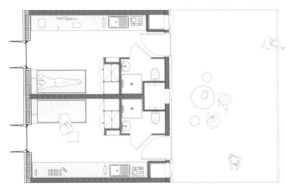

单人间单元
single room unit

夫妻间单元
couple room unit

0 1 2m

大田大学惠化寄宿学院
Daejeon University Hyehwa Residential College

Seung, H-Sang

森林中的学术修道院

韩国建筑师承孝相与大田大学之间的合作关系始于他2000年设计惠化文化中心之时。此后他继续领衔设计大田大学的其他项目,包括30周年纪念馆,馆中包括演讲厅和其他一些设施,其在校园中的地位宛如皇冠上的一颗宝石。最初,Min Hyun-sik教授的Kiohun建筑师事务所曾制定了一项对整个校园进行重组的总体规划,规划对象也包括宿舍和一个体育馆,以期彻底改变整所大学的面貌。得益于此次大整修,大田大学如今已成为当地一处颇受欢迎的人群聚集地。但在这些改变之前,大学东南部到处都充斥着老建筑,人们称之为"老城",而大多数新建筑则都建在西部。当时,大学为新宿舍选择了两个地址——一个是西部,也就是旧宿舍的所在地,另一个位于东边一个看起来并不适合动工的树木茂密的斜坡上。然而,这位建筑师表示他更喜欢后者——一则是因为该地区有必要建造新的建筑,二则是他有极大的热情来完成为这片树木茂密的斜坡设计全新建筑的挑战。

大学与职业学校之间的不同之处在于它不仅教授技能,还是一个培养全球普遍价值观的机构。这就要求它采取全面的教育方法,而不仅仅是要求学生通过死记硬背来获取知识。这就是为什么寄宿制度如此有效的原因——寄宿制学校的学生尽管只是暂时居住在此,但他们却仿佛是进入了另一个境界,就像僧侣们为了寻求身体、物质和精神上的自由而遁世一般。因此,寄宿制学校的开办理念和修道院的理念类似,该设施也因此被命名为"学术修道院"。建筑的墙体将校园与城市分隔开来,就像寺院把僧侣们与外界分隔开一样。寄宿学院综合体建

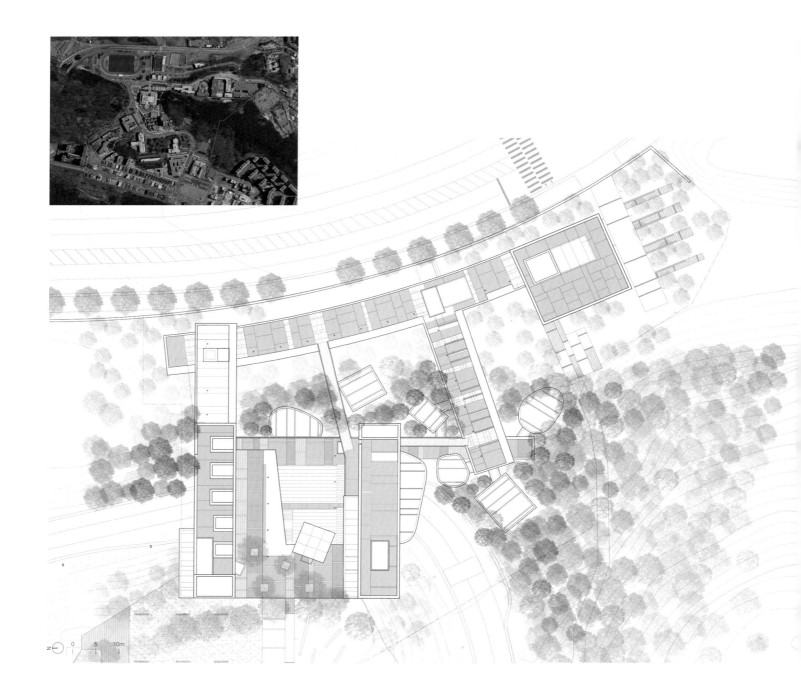

筑的内部是开放的，但其所处的孤立地势却遮住了来自外部世界的视线。建筑本身就构成了分界线，一些共用空间分布在其周围。这一布局之间来来往往的学生代表了生命的活力。在早期规划阶段，建筑师仔细地将建筑物设计在树木之间，该工程也因此被称为"森林中的学术修道院"。但是，由于树木被砍伐用于施工目的，因此大部分森林都消失了。尽管建筑师对这一行为颇为不满，但也做出了让步，条件是要在工地上再次种植新的树木。不过到目前为止，工地上只种了不多的树苗，这意味着这个特定的建筑项目尚未竣工。这位建筑师仍在憧憬着自己的愿景得以实现，盼望着学生们能够在这所满是葱郁树木的寄宿制学院中学习。

Academic Monastery in Forest

Korean architect Seung H-Sang's relationship with Daejeon University began in 2000, when he designed the Hyehwa Culture Center. He went on to lead other projects for Daejeon University including the 30th Anniversary Memorial Hall, which is the crown jewel of the campus, hosting lecture halls and other facilities. From the outset, Professor Min Hyun-sik's Kiohun Architect and Associates drew up a master plan to reorganize the entire campus – including

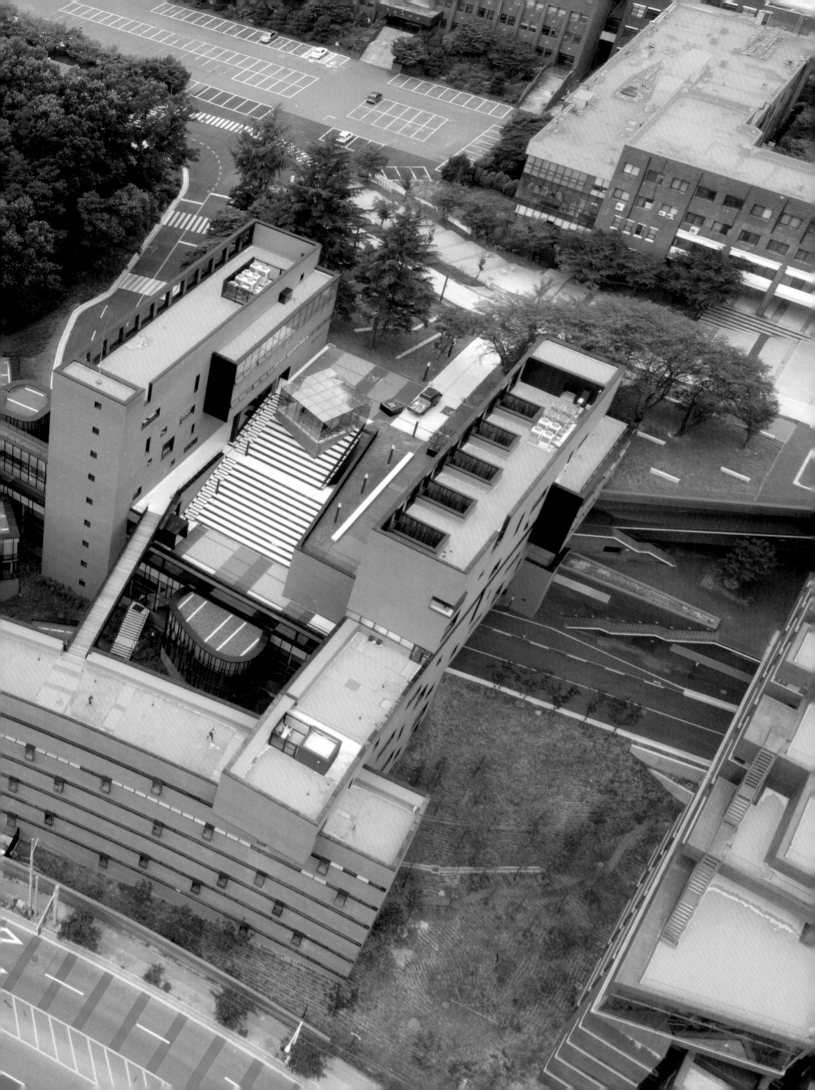

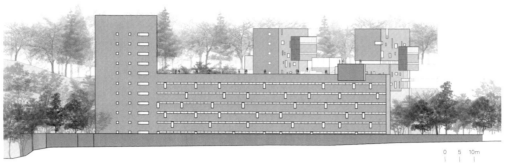
东立面 east elevation

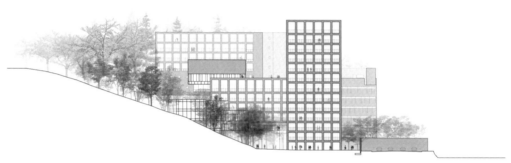
南立面 south elevation

西立面 west elevation

北立面 north elevation

dorms and a gymnasium – to completely alter the look of the university. Daejeon University is now a popular local hangout thanks to the overhaul, but prior to the changes, the area – southeast of the university, where the older buildings were located – was filled with old buildings and was known as the "old town" while most of the new buildings were placed in the west. At the time, the university identified two potential sites for the new dorms – one in the west, where the old dorms were, and the other on a wooded slope in the east that seemed ill-suited for architecture. The architect, however, voiced his preference for the latter site – because of the necessity of new buildings in the area, and his desire for the challenge of designing entirely new kinds of structures for the wooded incline.

A university differs from vocational schools in that it teaches more than skills – it is an institution that fosters global, universal values. This calls for holistic styles of education that go beyond the rote acquisition of knowledge. This is what makes the Residential College (RC) system so effective – students in the RC, though their stay is temporary, step into another world like monks who have renounced the world in search of physical, material, and spiritual freedom. For this reason, the RC's driving concept was that of a monastery, and the facility was named the Academic Monastery. It is a walled structure that separates the campus from the city, as monasteries do for monks. The RC complex is open internally, but its isolated topography veils the building from outside view. The residential buildings are themselves the boundaries, and shared spaces are dotted around the area. The dynamic flow of students through this layout will act as representation of the dynamism of life. In the early planning stages, the buildings were carefully positioned between the trees, and the project was called the Academic Monastery in the Woods. But much of the forest disappeared when the trees were cut down for construction purposes. Though dissatisfied by the act, the architect tolerated it under the condition that new trees would be planted on the site. Only a few saplings have been planted thus far, which means that this particular architectural project remains yet unfinished. The architect continues to wait for his vision of students studying among the full-grown trees at the RC to be realized.

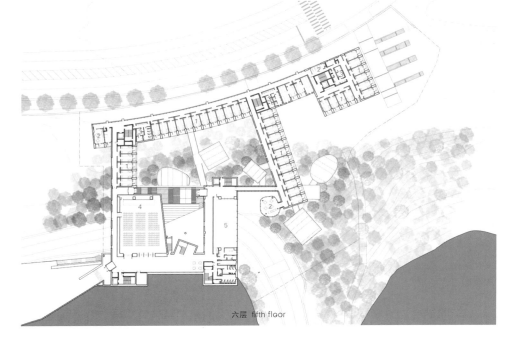

1. 宿舍
2. 公共休息室
3. 公共浴室
4. 多功能厅
5. 瑜伽和舞蹈室

1. dormitory
2. common room
3. communal bathroom
4. multipurpose hall
5. Yoga & dance studio

六层 fifth floor

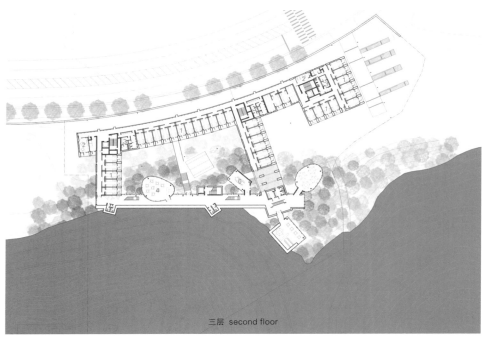

1. 宿舍
2. 公共休息室
3. 公共浴室
4. 学习室
5. 研讨室

1. dormitory
2. common room
3. communal bathroom
4. study lounge
5. seminar room

三层 second floor

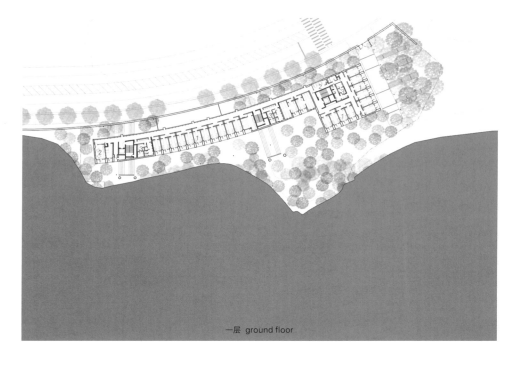

1. 宿舍
2. 公共休息室
3. 公共浴室

1. dormitory
2. common room
3. communal bathroom

一层 ground floor

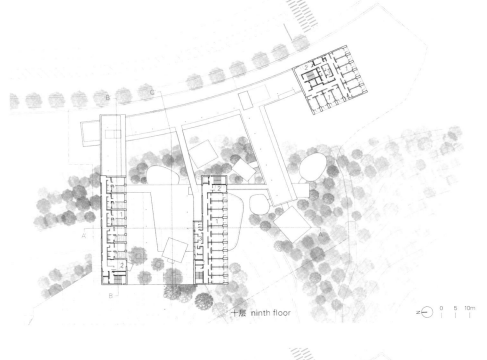

1. 宿舍
2. 公共休息室
3. 公共浴室

1. dormitory
2. common room
3. communal bathroom

十层 ninth floor

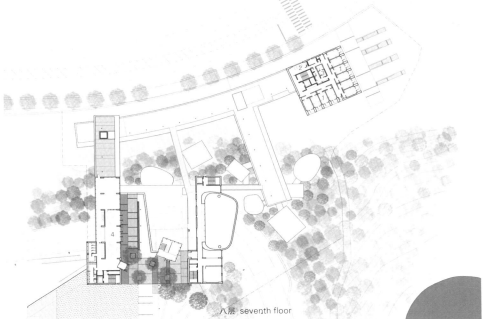

1. 宿舍
2. 公共休息室
3. 公共浴室
4. 信息中心

1. dormitory
2. common room
3. communal bathroom
4. information center

八层 seventh floor

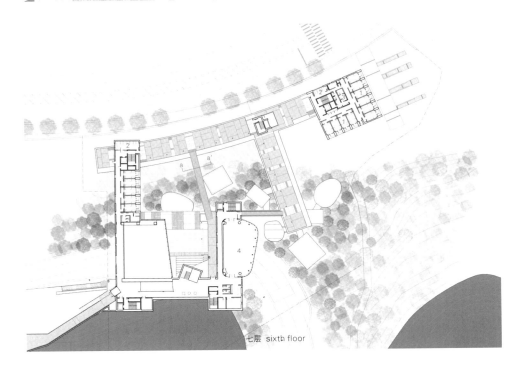

1. 宿舍
2. 公共休息室
3. 公共浴室
4. 健身中心

1. dormitory
2. common room
3. communal bathroom
4. fitness center

七层 sixth floor

1. 宿舍
2. 公共浴室
3. 健身中心
4. 瑜伽和舞蹈室
5. 多功能厅

1. dormitory
2. communal bathroom
3. fitness center
4. Yoga & dance studio
5. multipurpose hall

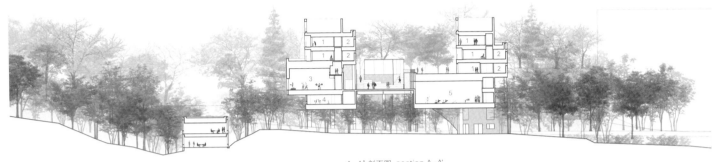

A-A' 剖面图 section A-A'

1. 公共休息室
2. 多功能厅
3. 信息中心

1. common room
2. multipurpose hall
3. information center

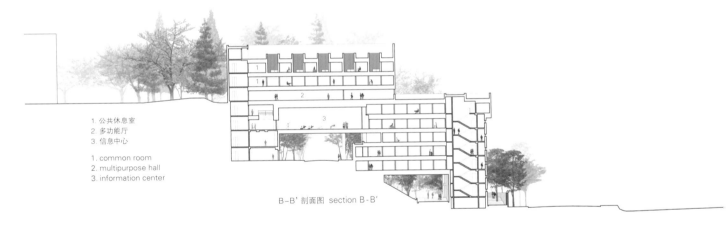

B-B' 剖面图 section B-B'

1. 多功能厅
2. 学习室
3. 宿舍

1. multipurpose hall
2. study lounge
3. dormitory

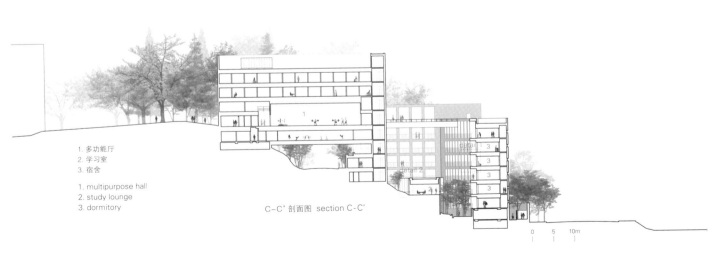

C-C' 剖面图 section C-C'

项目名称：Daejeon University Hyehwa Residential College
地点：Daejeon University, Daejeon, Korea
建筑设计：Seung H-Sang, Lee Dong-Soo, Kim Sung-Hee
结构工程：The Naeun Structural Engineering co.,ltd.
机械工程：DE TECH co.,ltd.
电气工程：Dae-Kyong Electrical Engineering & Consulting co.,ltd.
项目：dormitory, BDH room, seminar room, study lounge, community lounge, fitness center, yoga & dance studio etc.
建筑面积：3,913m²
总楼面面积：14,084.5m²
高度：45.6m
建筑规模：1 story below ground, 11 stories above ground
结构：reinforced concrete
外部材料：dressed brick, aluminium sheet, exposed mass concrete, thk 24 low-E double glazing glass
设计时间：2013.11—2015.2
施工时间：2015.5—2017.10
摄影师：©Kim Jong-oh (courtesy of the architect)

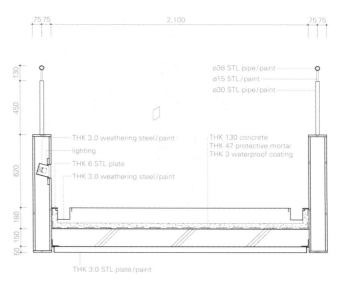

详图a-a' detail a-a'

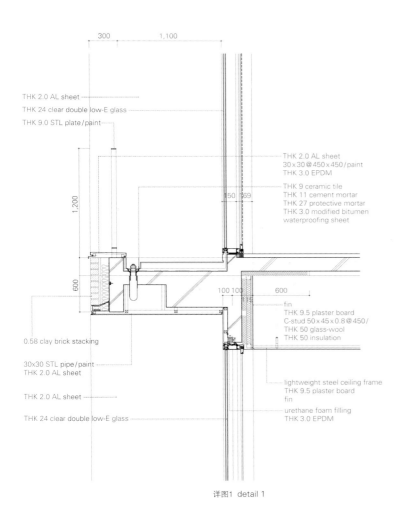

详图1 detail 1

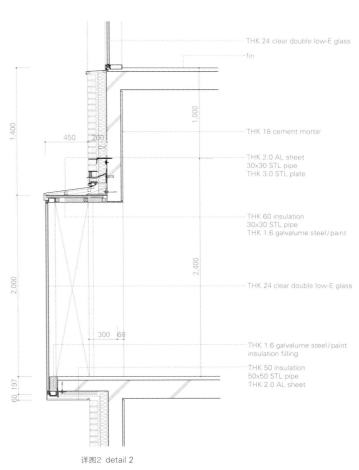

详图2 detail 2

学生宿舍——非类型化建筑 Student Housings – Against the Typology

塔帕尔大学学生宿舍
Thapar University Student Accommodation

McCullough Mulvin Architects

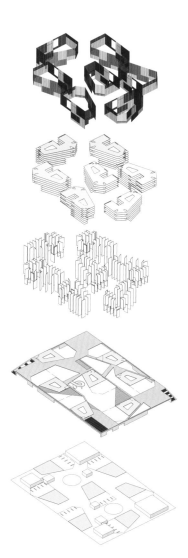

总部位于都柏林的McCullough Mulvin建筑师事务所与DPA合作，为这块101ha的场地设计了一份总体规划，来为塔帕尔大学打造一系列标志性建筑，工程包括分两个阶段建设的学生宿舍、一个体育中心和一个包括图书馆、讲堂和计算机科学大楼在内的学习中心。整个校园都采用了现代设计方法和材料，将大学定位为现代化的当代印度的一部分。

总体规划的一期工程包括一系列可容纳1200名学生的学生宿舍。L形塔楼是一座由单一材料建成的几何形状实体建筑。四个完工的单元外面覆盖的是与当地环境中的阿格拉红砂岩相匹配的光滑红宝石颜色，这四座塔楼共同形成了一个几何形状，它们在同一个矩形布局内却面向不同的方向。

四座塔楼由一个钢-混凝土结构基座连接起来，基座内容纳了接

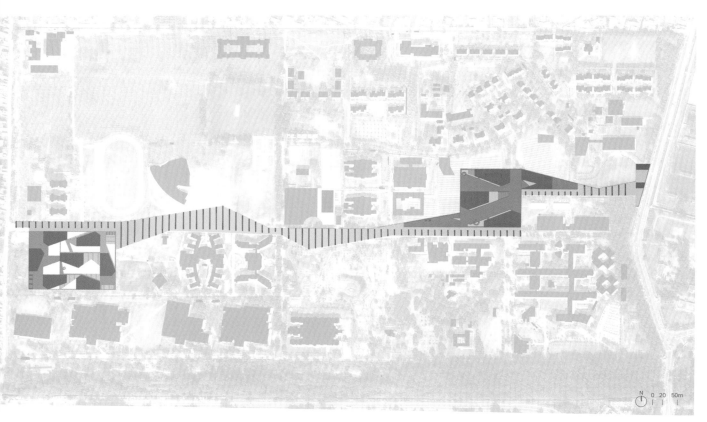

待处、健身房和餐厅。基座两侧分别有一部大型楼梯从地面通往上层。一系列走道让学生和工作人员有机会以一种全新的方式穿过帕蒂拉校区。一条有顶走道使人们在享受大自然所有便利设施的同时,还不会受到天气变化的影响。

McCullough Mulvin对塔帕尔大学校园内的建筑采取的做法是,将整个校园视为一个景观,并将建筑打造成一种新的自然地理格局,通过四通八达的走道,将部分建筑形式延伸出去,唤起人们对岩石高地和幽暗山谷的遐想。建筑通过飘浮基座、有顶走道和有围墙的花园提供降温和遮阳的作用。这是设计中的关键可持续要素,也是对印度建筑模式的参考。

二期工程将于2019年结束。总体规划也将于2020年全面竣工。

Dublin based practice McCullough Mulvin Architects in partnership with Design Plus Associates (DPA) have designed an overall masterplan for the 101ha site to provide a series of iconic buildings for Thapar University, including two phases of student residences, a sports center, and a new learning center incorporating a library, lecture theatres, and computer science building. A contemporary approach to design and materials has been employed across the campus to position the University as part of a modern, contemporary India.

The first phase of the masterplan is a series of student accommodation buildings providing residence for 1,200 students. The L-shaped towers provide an architecture of solid geometric forms made with single material. The four completed units are covered externally with red jali screen colour matching to the Agra Red Sandstone native to the local context, and together the towers make their own geometry, facing in different directions within a rectangular field of play. The towers are linked by a steel and concrete podium which shields the reception, gym and dining spaces below. A large staircase leads from the ground at either end and a series of walkways offer students and staff the opportunity to progress through the Patiala campus in a radically new way, protected from the weather with a covered walkway while enjoying all the amenities of nature.

McCullough Mulvin's approach to the architecture of Thapar University was to consider the whole campus as a landscape and to make a new natural geography out of the buildings, extending part of their built forms to evoke rocky heights and shaded valleys, with connecting walkways. The provision of cooling and shade through floating podiums, covered walkways and walled gardens are key sustainable elements of the design and are a reference to Indian models of architecture.

The second phase of the development will wrap up in 2019 with the masterplan to fully complete in 2020.

项目名称：Thapar University Student Accommodation / 地点：Patiala, Punjab, India / 事务所：McCullough Mulvin, Design Plus Associates / 工料测量师：Vinod Markanda, Delhi
承包商：Gannon Dunkerley & Co. Ltd. / 室内设计：Contractor – A. N. J., Mumbai / 结构工程师：Pristine Solutions / 机电工程师：Aeon Integrated Building Design
顾问 / 景观建筑师：Integral Landscape / 立面顾问：K. R. Sureish, Axis Façade Consulting / 客户：Thapar University
建筑面积：phase 1 – 38,459m², phase 2 – 29,390m² / 竣工时间：2018 / 摄影师：©Christian Richters (courtesy of the architect)

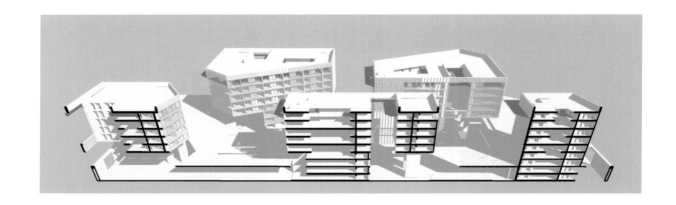

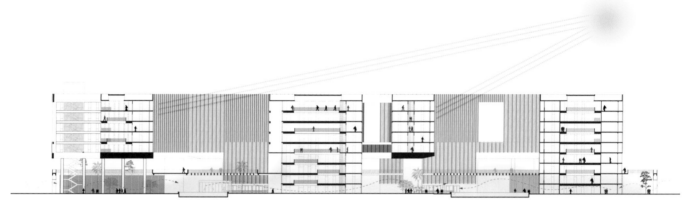

A-A' 剖面图 section A-A'

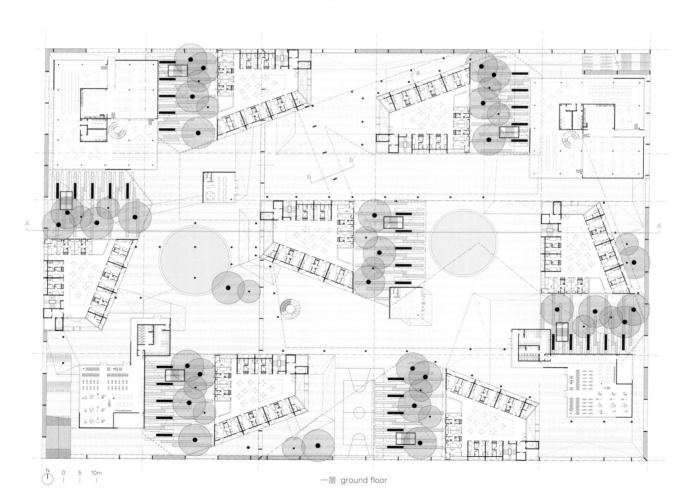

一层 ground floor

三层 second floor

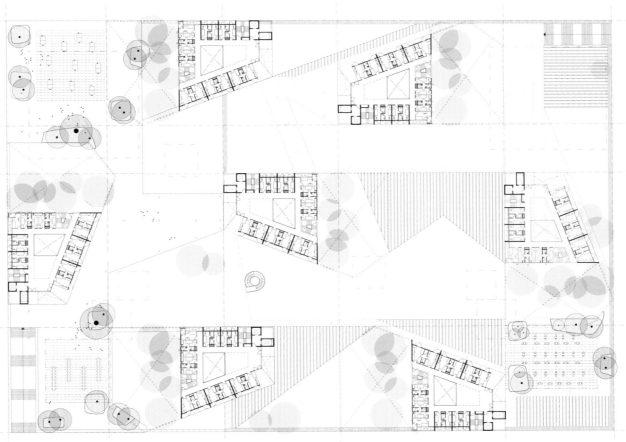

二层 first floor

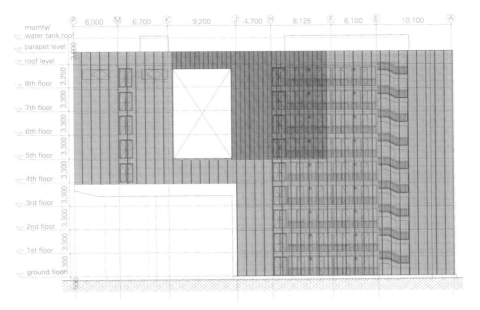

B-B' 立面图 elevation B-B'

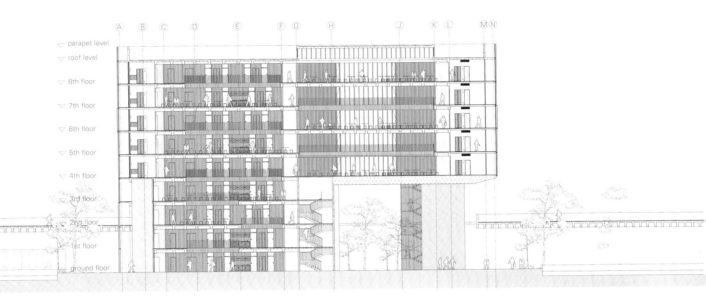

C-C' 剖面图 section C-C'

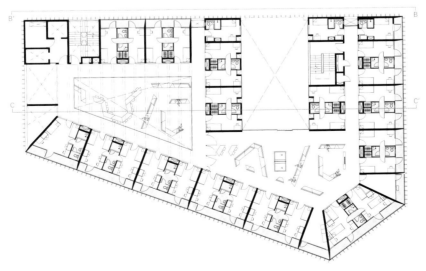

招待所楼层 hostel unit floor

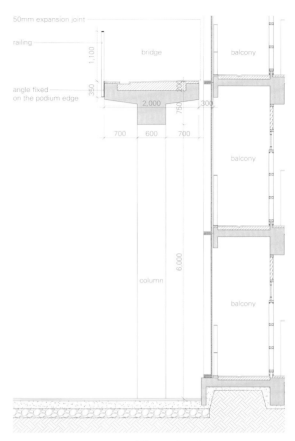

a-a' 详图 detail a-a'

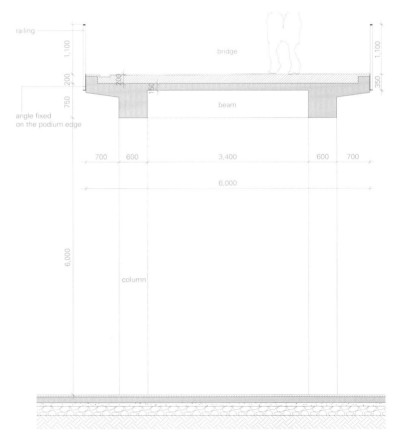

b-b' 详图 detail b-b'

城西国际大学爱家宿舍
Josai International University iHouse Dormitory

Studio Sumo

日本人口逐渐减少,日本的大学试图吸引越来越多的国际学生长期或短期留校。本案宿舍和国际中心能为大约140名国际学生提供有效的住房和教育场地,通过共享公共空间和房间,将文化背景不同、经济条件不同的人融合在一起。这些共享空间和房间包括带私人浴室的单人间和双人间,以及有共用洗浴设施的四人间。城西国际大学将这些共用房间以每月仅80美元的价格租给学生住宿,这一政策主要是针对来自亚洲和东欧发展中国家的中等收入学生。

这座建筑坐落在通往大学的主干道上,处在一片稻田的边缘。建筑包括一个9m宽的宿舍酒吧,酒吧悬浮在国际中心上空,而国际中心向外延伸与周围景观融为一体。国际中心由一间画廊、一个档案室和一个纪念已故宪仁亲王的活动空间组成,他帮助推动日韩双方合作举办了2002年世界杯。城西国际大学与宪仁家族一直保持着密切联系。

要想进入建筑,需要离开校园主楼,穿过酒吧的空隙空间,这个空隙空间将建筑地面层分成两个部分。建筑采用百叶状表面设计,其中伸出多个阳台,百叶状表面遮住了室外走道,这些室外走道服务于朝向外面稻田的宿舍房间。建筑表面由不同尺寸的成品铝百叶组成,每个百叶都最大限度向外延伸,且有一个竖直支架制成。无论是在立面图还是在剖面图上看都形成了互相交织的结构。设计百叶的目的是为宿舍楼遮阳,创造一个整齐划一的立面,以反映统一的身份特点,而不仅仅是一个个宿舍单元的简单集合。朝南的铝百叶白天随着太阳轨迹而改变外表颜色,使建筑物外表从白色变成银色,日落时变成橙色。百叶立面的后边面向走道设置多扇推拉玻璃门,能让人想到日本传统住宅中的"掾侧(连接传统和室和户外空间的木结构走廊)"空间。共享空间、走道和阳台系统将拥挤的生活空间扩展到了户外。

As Japan copes with a declining population, universities are trying to attract an increasingly international student body for both long and short term stays. This dormitory and International Center for approximately 140 international students efficiently houses, educates, and integrates a population that is both culturally and economically diverse through a collection of shared public spaces and rooms that range from singles and doubles with private baths, to rooms for four persons with shared bathing facilities. These shared rooms allow Josai International University to offer accommodations for as little as $80/month and are particularly geared towards students of modest means from emerging Asian and Eastern European countries.

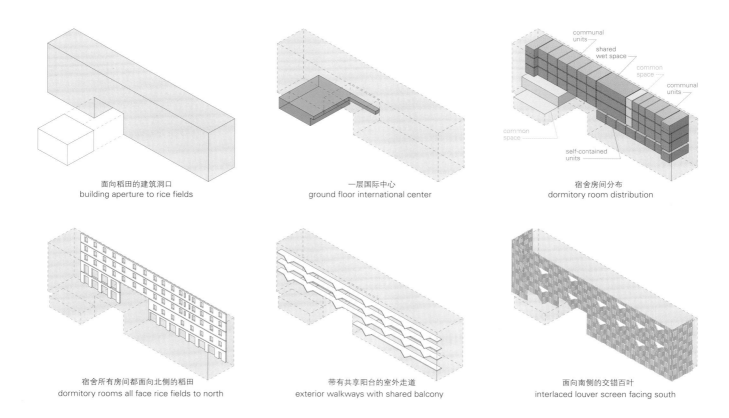

面向稻田的建筑洞口
building aperture to rice fields

一层国际中心
ground floor international center

宿舍房间分布
dormitory room distribution

宿舍所有房间都面向北侧的稻田
dormitory rooms all face rice fields to north

带有共享阳台的室外走道
exterior walkways with shared balcony

面向南侧的交错百叶
interlaced louver screen facing south

1. 城西国际大学爱家宿舍
2. 原有澄清池/圆形剧场
3. 宪仁亲王纪念地
4. 进入主校园的入口
5. 入口道路
6. 稻田
7. 原有储藏室
8. 行李箱和自行车存放处
9. 原有停车场

1. JIU iHouse Dormitory
2. existing retention pond/ampitheater
3. Prince Takamado memorial field
4. entry to main campus
5. access road
6. rice fields
7. existing field house
8. trunk and bicycle storage
9. existing parking

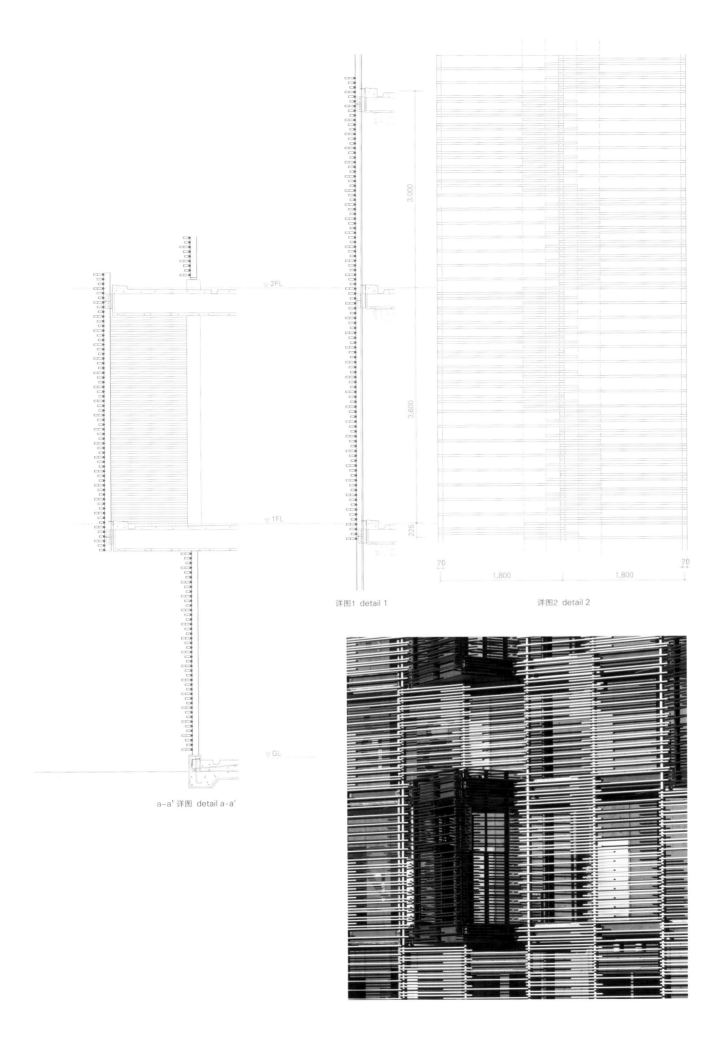

详图1 detail 1

详图2 detail 2

a-a' 详图 detail a-a'

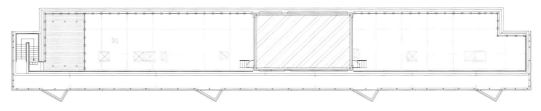

屋顶 roof

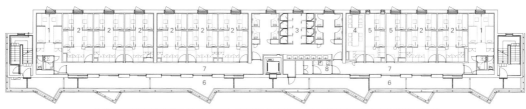

1. 带卫浴间的双人间宿舍 2. 四人间宿舍 3. 共用浴室 4. 咖啡厅 5. 双人间宿舍 6. 室外走道 7. 走廊 8. 餐具室
1. double dorm room w/bathroom 2. four persons dorm room 3. shared bathroom 4. coffee room
5. double dorm room 6. exterior walkway 7. hallway 8. pantry

三层/四层/五层
second/third/fourth floor

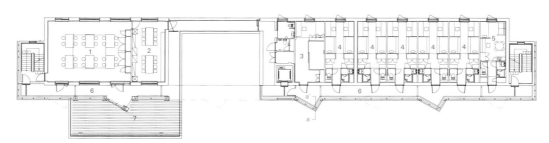

1. 公共房间 2. 小组厨房 3. 无障碍宿舍 4. 带卫浴间的双人间宿舍 5. 客座教师房间 6. 室外走道 7. 露台
1. common room 2. group kitchen 3. wheelchair accessible dorm room 4. double dorm room w/bathroom
5. visiting faculty room 6. exterior walkway 7. terrace

二层 first floor

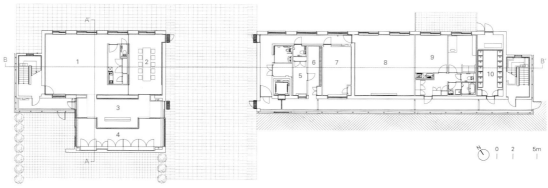

1. 国际中心活动室 2. 国际中心档案室 3. 国际中心接待处 4. 宪仁亲王纪念画廊 5. 宿舍门岗
6. 收发室 7. 研讨室 8. 档案室 9. 住宿教职员工公寓 10. 洗衣房
1. international center event room 2. international center archive room 3. international center reception
4. prince takamado memorial gallery 5. dormitory guard booth 6. mailroom 7. seminar room
8. activity room 9. resident faculty apartment 10. laundry

一层 ground floor

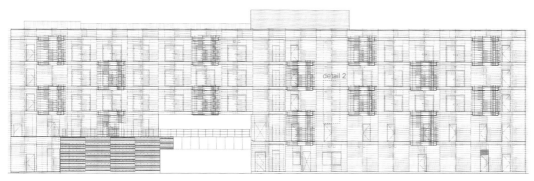

西南立面 south-west elevation

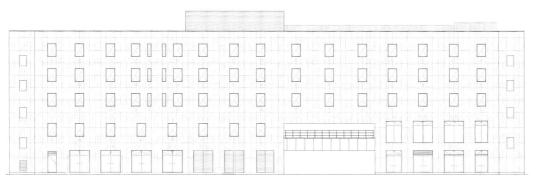

东北立面 north-east elevation

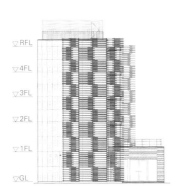

西北立面 north-west elevation

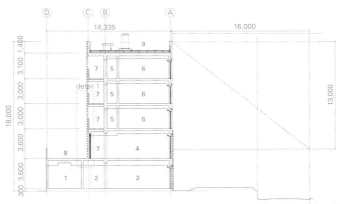

1. 宪仁亲王纪念画廊 2. 国际中心接待处 3. 国际中心活动室 4. 公共房间 5. 大厅 6. 宿舍 7. 走道 8. 露台 9. 屋顶平台
1. prince takamado memorial gallery 2. international center reception 3. international center event room
4. common room 5. hall 6. dorm room 7. walkway 8. terrace 9. roof deck

A-A' 剖面图 section A-A'

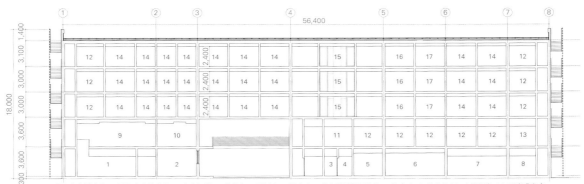

1. 国际中心活动室 2. 国际中心档案室 3. 宿舍门岗 4. 收发室 5. 研讨室 6. 活动室 7. 住宿教职员工公寓 8. 洗衣房 9. 公共室 10. 小组厨房 11. 无障碍宿舍
12. 带卫浴间的双人间宿舍 13. 客座教师房间 14. 四人间宿舍 15. 共用浴室 16. 咖啡厅 17. 双人间宿舍

1. international center event room 2. international center archive room 3. dormitory guard booth 4. mailroom 5. seminar room 6. activity room
7. resident faculty apartment 8. laundry 9. common room 10. group kitchen 11. wheelchair accessible dorm room
12. double dorm room w/ bathroom 13. visiting faculty room 14. four persons dorm room 15. shared bathroom 16. coffee room 17. double dorm room

B-B' 剖面图 section B-B'

The building is sited along the main access road to the university on the edge of an expanse of rice fields. It is comprised of a 9-meter wide dormitory bar that hovers over the International Center that projects out to engage the landscape. The International Center is comprised of a gallery, an archive room, and an event space for remembering the late Prince Takamado who helped to promote the Japan/Korea partnership that hosted the 2002 World Cup. JIU maintains a close relationship with the Takamado family.

One enters the building off the campus road through a void in the bar that separates the two programs on the ground level. A louvered surface interspersed with projecting balconies masks exterior walkways that serve the dormitory rooms facing out over the rice fields beyond.

This surface is comprised of off-the-shelf aluminium louvers of different dimensions, each cantilevering to its maximum allowable extension past a vertical support. This creates an interwoven texture in both elevation and section. The goal of the louver was to mask the dormitory program and create a unified facade that reflected a singular identity rather than a collection of units. The south-facing aluminium tracks the sun over the day, turning the building from white, to silver, to orange at sunset. Behind the louvered facade, multiple sliding glass doors open onto the walkways, recalling the "engawa" space of traditional Japanese houses. The provision of shared spaces, the walkway, and the balcony system expand the compressed living space into the outdoors.

项目名称：iHouse Dormitory
地点：Togane-shi, Chiba-ken, Japan
建筑师：Studio SUMO – Sunil Bald, Yolande Daniels (Partners in charge), Edward Yujoung Kim, Teo Quintana, Kim Jae-hyun, Kevin Sani, Yezi Dai, Oh Young-tack, Masahi Takazawa (Project team)
记录建筑师：Obayashi Corporation – Koji Onishi (General manager), Toshimichi Takei (manager), Atsuko Mori (Deputy manager), Mao Shigeishi, Hikaru Takei
结构、土木、岩土工程师，施工经理，总承包商，景观建筑师，施工监理：Obayashi Corp.
照明设计：Studio SUMO, Obayashi Corp.
客户：Josai University Educational Corp.
总楼面面积：2,800m²
造价：USD 15million
竣工时间：2016
摄影师：©Kawasami Kobayashi Photograph Office (courtesy of the architect) - p.175, p.176, p.177, p.180, p.181, p.182, p.185 top; ©Kudoh Photography Ltd. (courtesy of the architect) - p.172~173, p.183, p.184, p.185 bottom-left, bottom-right

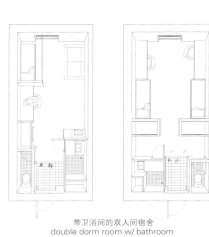

带卫浴间的双人间宿舍
double dorm room w/ bathroom

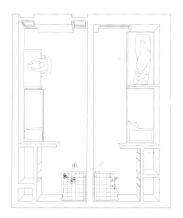

共用浴室
shared bathroom

双人间宿舍
double dorm room

卢西安·考尼尔学生宿舍
Lucien Cornil Student Residence

A+Architecture

八层木结构建筑

A+Architectcture建筑事务所在马赛为法国学生服务中心设计了法国最高的木结构建筑之———卢西安·考尼尔学生宿舍。

这座八层高的学生宿舍楼是一个成功的环境塑造项目的结果。其巧妙的城市化设计方法使这座有着200个房间的结构成为一座功能性建筑，舒适且面向城市开放。

该项目包括三座翼楼，设计得益于高高的一层、最上面两层的阁楼，以及高质量的共享空间。该项目的高度提高了，可以与周围建筑相互影响，使这些建筑在密度如此大的区域也有呼吸空间。

大多数房间都朝向封闭的花园，这是一个真正令人感到放松的室内花园，坐落于街道一侧，花园的洞口沿着一条噪声较小的小巷而设。在这个狭小的城市环境中，到处可见木结构建筑。为了减少工程对正常使用的影响，建筑师提出了优化的施工日程安排，承诺宿舍的舒适度，所有这些原因使法国学生服务中心坚信这一设计探索定能成功。

所有的天花板和房间的墙体都采用了木材，且墙体设置了隔声。走廊和公共房间也是如此，但老化太明显的墙面则不适用。硬朗的室内结构却有着温暖和放松的氛围，而且室内的声效也非常柔和。木框架采用交叉层压的组装方式，散发出森林的气息。交叉层压实木的使用减少了能源消耗，而且在减少碳足迹方面达到了卓越的水平。

A+Architecture建筑事务所道："整座建筑的设计，在保持线条一致和价格具有竞争力的同时，还保证了良好的保温隔声效果。"建筑覆层的设计也很重要。穿孔曲面板与大块铝板混合使用，使线条融为一体，减小了建筑规模，也减少了建筑体量的数量。

多孔的建筑表皮从一部分宽宽的带状玻璃窗前经过，灯光透过百叶，把傍晚的建筑变成了马赛夜晚的灯塔。经过景观美化打造的室内花园主要面向城市，可承办会议；大广场连接着圣皮埃尔街入口，入口处一棵精心保存的雄伟松树成为这里的亮点。所有地方的光线都十分充足，有些地方的光线经过后面覆层保护性盖板孔洞的过滤，有些地方的光线是从公共区域上部铝栏杆后面发出来的，这样的设计对宿舍房间来说有点奢侈。多孔表皮缺失的地方安装了卷帘，放下卷帘就可以关闭整个洞口。

在这个建筑密度较大的地区，该项目的选址和空间选择却实现了公共区域、交通流线区域的良好设计，而且在建筑中还可以欣赏到城市的景色。

木结构与设计巧妙且实用的建筑结合在一起，提供了一个非常合时宜的解决方案，不但实现了创新，还非常好地适应了环境。

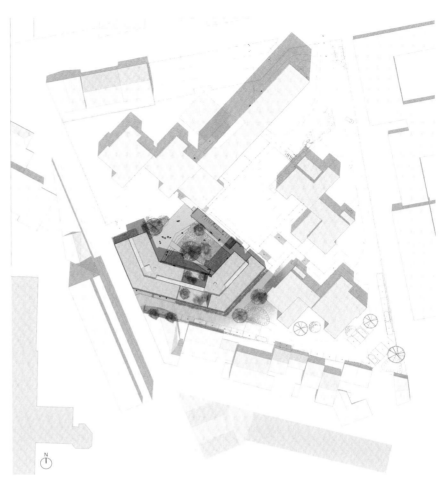

项目U形布局
U-shaped layout of the project

原理周围建筑，转向城市
moving away from the surrounding buildings; turning towards the city

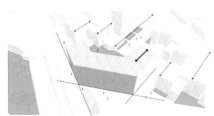

重新发现城市的原始面貌
rediscover the original aspect of the city

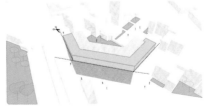

雕刻出可以看到周围建筑的阁楼
carve out an attic over the surrounding buildings

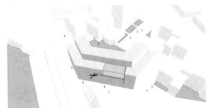

在南立面创造呼吸空间
create breathing space on the south frontage

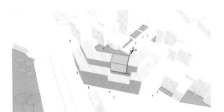

在北立面设立一个中断空间
create a break in the north frontage

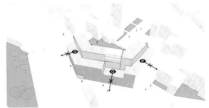

建筑师没有忽视交通流线空间
circulation areas that are not overlooked

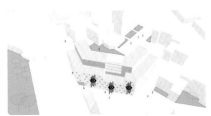

覆层的设计使宿舍空间变得与人更加亲密
cladding that filters to make the accommodation more intimate

Eight-Floor Building with Wooden Structure

In Marseilles, A + Architecture has designed one of the highest wooden buildings in France for the CROUS: the Lucien Cornil hall of residence.

This eight-floor student residence is the fruit of a successful environmental construction project. Its sensitive urban approach makes this 200-room structure a functional building, comfortable and opening out towards the city.

Consisting of three wings, the design benefits from a very high ground floor and attics on the top two levels as well as shared spaces of high quality. The graduation of the building heights of the project will interact with the surrounding buildings and leave them with space to breath despite the density of the area.

Most of the rooms are directed towards the enclosed garden, a genuinely relaxing indoor garden, on the street side, whose openings are positioned along the less noisy alley.

In this constricted urban environment, the choice of wood construction was obvious. Reduction in disruption caused by the works, an optimised schedule, and also a commitment to the comfort of the residence are what convinced the CROUS to embark on the adventure.

Wood is found on all the ceilings and the walls of the rooms, the latter being sound-proofed. It is also present in the cor-

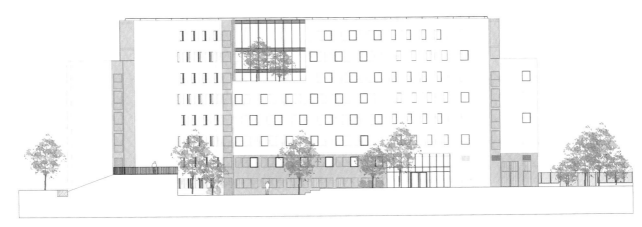

北立面 north elevation

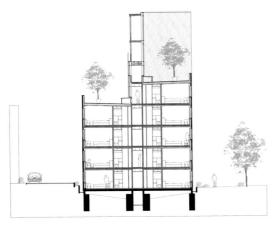

A-A' 剖面图 section A-A'

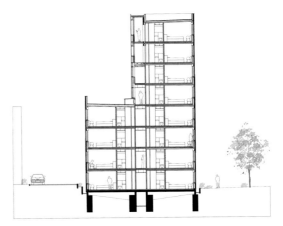

B-B' 剖面图 section B-B'

项目名称：Lucien Cornil Student Residence / 地点：Marseilles, Bouches-Du-Rhône, France / 事务所：A + Architecture / 调查组织：Alpes Contrôle
安全协调：QUALICONSULT / 结构设计：TPFI / 声学设计：Celsius Environnement / 环境设计：L'ECHO
经济学家：OPC：ARTEBA / 总承包商：Travaux du Midi (Vinci) / 木结构方案设计与建造：Arbonis (Vinci) / 客户：Crous in Aix-Marseille Avignon
方案：wooden construction of 200 lodgings in an eight-floor building, Cité U Lucien Cornil site / 用地面积：12,000m² / 建筑面积：4,352m² / 造价：EUR 8,715,000
设计时间：2015 / 竣工时间：2017.9 / 摄影师：©Benoit Wehrle (courtesy of the architect)

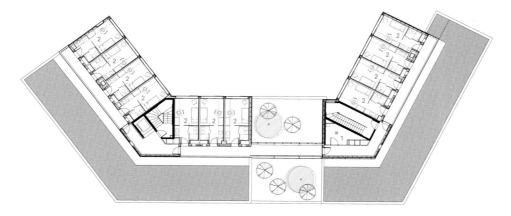

1. 家庭房间
2. 宿舍房间

1. household room
2. room

六层 fifth floor

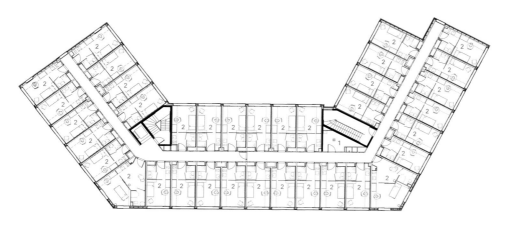

1. 家庭房间
2. 宿舍房间

1. household room
2. room

二层 first floor

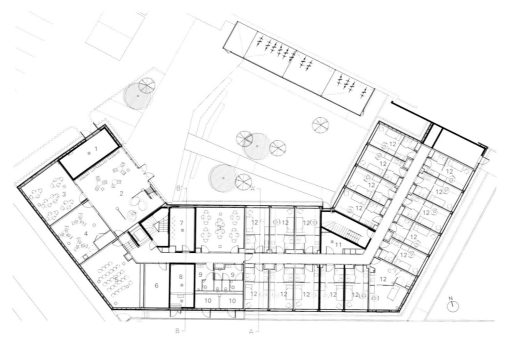

1. 家庭房间垃圾房
2. 接待大厅
3. 学习室
4. 公共房间
5. 电视/视频播放室
6. 健身房
7. 洗衣房
8. 下车点
9. 卫生间
10. 技术室
11. 家庭房间
12. 宿舍房间

1. household garbage
2. reception hall
3. study room
4. common room
5. TV/video room
6. gym
7. laundry
8. under station
9. toilet
10. technical room
11. household room
12. room

一层 ground floor

ridors and communal rooms, but not on the facings where its ageing is deemed too visible. Its strong interior presence gives the impression of a warm and relaxing atmosphere with soft acoustics. The wooden shrouds, with cross-laminated assembly, give off a forest scent. The use of solid wood CLT (Cross Laminated Timber) limits energy consumption and provides an excellent carbon footprint.

"The entire building has been designed to be very thermally and acoustically efficient, while maintaining consistent lines and at a very competitive price", states A+Architecture.

The cladding is something else. A perforated curved panel is mixed with large aluminium shingles to mix up the lines, reduce the scale and breakdown the volumes.

The perforated skin passes in front of a section of wide glazed strips, transforming the building in the evening to a beacon of light in the Marseillaise night. The landscaped interior garden, mainly pointing towards the city, is given over to meetings, the large piazza connecting with the rue Saint Pierre entrance that highlights the carefully preserved majestic pine tree. The light is magnified everywhere, sometimes filtered behind the perforations of the cladding's protective cladding sheets or behind the aluminium railings in the upper sections of the communal areas, somewhat generous for the rooms; the blanking is provided by a roller shutter which closes off the entire opening.

In a dense area, the location and choice of space occupied have enabled the communal areas, the circulation areas and the views to the city.

The wooden structure combined with a sensitive and functional building provides a solution very much of its time; innovative and in-tune with the environment.

P172 Studio Sumo

Has been operated by Yolande Daniels[left] and Sunil Bald[right] since 1998, based in New York. Committed to architecture and design responsive to social and cultural contexts through the solutions that are unexpected, inventive, and engaging. Yolande Daniels received her BA in Architecture from City College of New York and M. Arch from Columbia University. Is currently Visiting Professor of Architecture at MIT. Sunil Bald received BA in Biology from the University of California, Santa Cruz and M Arch from Columbia University. Is Associate Professor and Associate Dean at the Yale School of Architecture. Was the recipient of the Annual Award in Architecture from the American Academy of Arts and Letters in 2015. Received the Design Vanguard Award from Architectural Record, the Emerging Voices and Young Architects Award from the Architectural League of New York.

P120 ppa architectures

Endeavours – particularly within the rapidly developing conurbation of Toulouse – to create urban fabric through buildings and urban projects whose varied programmes and contexts are systematically analysed and questioned from the point of view on both use and construction. Tries to adapt neutral form of flexible and frugal construction to specific architectural and urban proposals that are generous and comfortable. This ambitious and pragmatic intention is based on a collaborative way of working, multidisciplinary and open design, adapting to today's urban and architectural.

P120 Scalene Architectes

Was founded in 2008 by Luc and Jean Larnaudie. In ten years of practice, Scalene has gained international recognition with its spontaneous and precise projects. Promotes a crossover architecture of exchange, which leaves room for the unforeseen. It works with what Luc & Jean call signs and experiences, which are the elements of a generative method, a tool for action. They establish the outline of the projects. Further experiences on this informal canvas are then based on the five senses and potential uses. Jean & Luc have been honored with numerous national and international awards. Currently Scalene is working on several projects in Europe and Asia. In addition, teaching is a significant activity of the office.

P120 Almudever Fabrique d'Architecture

Was founded in 2003 by French architect, Joseph Almudever. He received his architecte DPLG in 1976 and a post-graduate diploma(DEA) of town planning and development in 1978 from the architecture school of Toulouse, Mirail. Started his career with Daniel Hermet and Christian Lefebvre in 1978 and established Almudever Lefebvre Agency. Was president of the Regional Council of Institute of Architects, Midi-Pyrénées in 1999-2003. Has been a professor of contemporary architecture at Toulouse Fine arts School since 1992.

P186 A+Architecture

Founded in Montpellier over 30 years ago, it is the largest architectural firm on the French shores of the Mediterranean, from Perpignan to Nice. A+ Group includes A+ Architecture, Arteba, l'Echo, Celsius Environnement and A+ Toulouse. Its team is organized by determined and creative 85 staffs including architects, urban planners, and engineers. The group specialises in large scale projects for public and private clients.
Philippe Cervantes, Gilles Gal, Philippe Bonon, Issis Raman, Christophe Aubailly, and Vincent Nogaret, from left, in the picture.

P136 Seung, H-Sang

Born in 1952, studied at Seoul National University. Worked for Kim Swoo-geun from 1974 to 1989 and established his office 'IROJE architects & planners' in 1989. Published several books including *Beauty of Poverty* (1996, Mikunsa), and taught at North London University, Seoul National University and Korea National University of Art, and TU Wien as visiting professor. The America Institute of Architects invested him with Honorary Fellow of AIA in 2002, and Korea National Museum of Contemporary of Art selected him as 'The Artist of Year 2002'. In 2007, Korean government honored him with 'Korea Award for Art and Culture', and he was director for Gwangju Design Biennale 2011 after for Korean Pavilion of Venice Biennale 2008. Was the first City Architect of Seoul Metropolitan Government in 2014 and finished its term in 2016. And now he is Chair Professor at Dong-A University and working as Chief Commissioner of Presidential Commission on Architecture Policy.

P156 McCullough Mulvin Architects

Dublin based practice McCullough Mulvin Architects is renowned for their award-winning work on cultural and civic buildings, particularly in education facilities throughout Ireland. Have been shortlisted for a prestigious World Architecture Award for the first phase of Thapar University in Northern India. Is motivated by contexts, layering architecture as a specific response to the site. Their approach to the architecture of Thapar University was to consider the whole campus as a landscape and to make a new natural geography out of the buildings, extending part of their built forms to evoke rocky heights and shaded valleys, with connecting walkways. Ronan O'Connor, Niall McCullough, Valerie Muvlin, Coran O'Connor, Ruth O'Herlihy, from left, in the picture.

P74 Max Núñez

Max Núñez Bancalari was born in Santiago, Chile in 1976 and studied at the Pontifical Catholic University of Chile, receiving his Degree and Masters in Architecture in 2004. Co-founded dRN Architects in 2005 with Nicolás del Río in Santiago, Chile. Received a Masters in Advanced Architectural Design in Columbia University, New York in 2010 where he received the Lucille Smyser Lowenfish Memorial Prize, and the William Ware Prize for Excellence in Design. Founded Max Nuñez Arquitectos in 2010 after 5 years at DRN Arquitectos. Is currently head of the Masters in Architecture Program at Pontifical Catholic University of Chile. Was nominated for the Mies Crown Hall Americas Prize 2014-2015 and 2016-2017, in the category Emerging Architecture. Received Design Vanguard from *Architectural Record* Magazine in 2017. Won the Best New Private House at Wallpaper Design Awards 2018.

P42 Pezo Von Ellrichshausen

Is an art and architecture studio founded in 2002 by Mauricio Pezo[left] and Sofia von Ellrichshausen[right]. They live and work in the southern Chilean city of Concepcion. They have taught at the Illinois Institute of Technology in Chicago, the Universidad Catolica de Chile in Santiago and Cornell University in Ithaca, New York. Among other venues, they have lectured at the Tate Modern, the Victoria & Albert Museum, the Metropolitan Museum of Art, the Alvar Aalto Symposium and the Royal Institute of British Architects. Their work has been distinguished with the Mies Crown Hall Americas Emerge Prize by the IIT, the Rice Design Alliance Prize, the Iberoamerican Architecture Biennial Award and the Chilean Architecture Biennial Award.

P94 Felipe Assadi Arquitectos

Felipe Assadi graduated as an architect from the Universidad Finis Terrae and earned a master's degree from the Pontificia Universidad Católica de Chile. In 1999, he won the 'Promoción Joven' prize of the Colegio de Arquitectos de Chile, awarded to the best architect under the age of thirty-five. He has taught at universities in Chile, Mexico, Brazil, Italy, Colombia, and the United States. Since 2011 he has been the Dean of the architecture school of the Universidad Finis Terrae. He has participated in exhibitions in Chicago, Barcelona, Pamplona, London, Quito, Tokyo, Venecia and Santiago, and his works have been constructed in Chile, Argentina, Uruguay, Mexico, Guatemala, Peru, Puerto Rico, Venezuela, the United States and Ecuador.